NUTRITIONAL AND METABOLIC
INFERTILITY IN THE COW

NUTRITIONAL AND METABOLIC INFERTILITY IN THE COW

T.J. McClure, BVSc, PhD, MACVSc, MRCVS

*Consultant in Cattle Health and Production,
Former Associate Professor in
Veterinary Medicine,
University of Sydney,
Australia*

CAB INTERNATIONAL

CAB INTERNATIONAL　　Tel: Wallingford (0491) 832111
Wallingford　　　　　　Telex: 847964 (COMAGG G)
Oxon OX10 8DE　　　　Telecom Gold/Dialcom: 84: CAU001
UK　　　　　　　　　　Fax: (0491) 833508

© CAB INTERNATIONAL 1994. All rights reserved.
No part of this publication may be reproduced in any
form or by any means, electronically, mechanically, by
photocopying, recording or otherwise, without the prior
permission of the copyright owners.

A catalogue entry for this book is available from the
British Library.

ISBN 0 85198 892 X

Typeset by Colset Pte Ltd, Singapore
Printed and bound in the UK

Contents

Preface	viii
1 INTRODUCTION	1
Mating Practices	2
The duration of the mating season	2
The control exerted over mating	4
The mating method	6
Standards of Reproductive Performance	6
Monitoring Reproduction	8
Intensively managed herds with seasonal calving	9
Intensively managed herds with all-year-round calving	14
Extensively managed herds	15
Reproductive Physiology	15
2 CATTLE NUTRITION	19
Feeding Practices	20
Pastures/crops	20
Compounded feedstuffs	20
Combination of concentrates and pastures/crops	21
By-products	21
Feed Intake	21
Appetite	21
Availability	22
Chemical composition	22
Feed Composition	22
Pasture and forage crops	23
Concentrates	26

Feed Type	27
Pasture	27
Hay and silage	28
Crops	28
By-products	29
Manufactured feedstuffs	29
Energy	30
Protein and Non-protein Nitrogen	31
Minerals and Vitamins	32
Nutrient Requirements	33
3 OCCURRENCE, PREVALENCE AND INCIDENCE OF NUTRITIONAL AND METABOLIC INFERTILITY	**34**
Occurrence	35
Prevalence and Incidence	38
4 AETIOLOGY, PATHOGENESIS AND CLINICAL SIGNS	**39**
Aetiology and Pathogenesis	39
Bodyweight and condition	40
Pregnancy, lactation and suckling	46
Minerals	47
Energy	50
Protein and non-protein nitrogen	57
Anti-oxidants	59
Clinical Signs	61
Energy, protein and minerals included in energy metabolism	61
Miscellaneous minerals and vitamins with a common anti-oxidant function	61
5 DIAGNOSIS	**62**
History	63
Clinical	63
Reproductive	63
Nutritional	73
Clinical Examination	82
Clinical Pathology	83
Differential Diagnosis	83
Controlled Therapeutic Trials	83
6 TREATMENT, CONTROL AND PREVENTION	**86**
Roughage	86
Concentrates	89
The Very High-producing Cow	89
Monitoring	92

Nutrition	92
Clinical signs	93
Clinical pathology	94
Heat Detection	94
Hormone Therapy	95
Weaning	95
Conclusions	95
References	97
Index	125

Preface

Nutritional and Metabolic Infertility in the Cow has been written for veterinary students, veterinary surgeons in cattle practice, agricultural science students specializing in animal husbandry and consultants in cattle management. I hope that the book will assist them in helping the managers of dairy and beef herds to control and prevent a significant part of the reproductive wastage that remains now that brucellosis, the venereal infectious diseases and male infertility have been largely controlled, eradicated or prevented. Research scientists may find this a useful introduction to studies on reproductive biology.

I wish to acknowledge the help of many farmers ard colleagues in my studies on nutritional and metabolic infertility in cows and in particular the help of S.L. Hignett, D. McFarlane, J.M. Payne, R.B. Heap, R.G. Dyer, Anne (Dowell) Ward, Sylvia (Park) Fry, H. Harris and R. McIntosh. Professors R.C. Kellaway, H. Lloyd-Davies and G. Stone kindly checked earlier drafts of the manuscript.

I thank my wife, Annette, and daughter, Dr Susan McClure, for their encouragement and assistance.

Nutritional and Metabolic Infertility in the Cow should be read in conjunction with modern reference books on bovine nutrition and reproductive physiology. Reference will need to be made to tables of nutrient requirements prepared for the local national herds and the composition of feedstuffs fed.

T.J. McClure
'Willunga'
Cassilis, New South Wales, Australia

1
INTRODUCTION

Nutritional diseases, or malnutrition, result from the ingestion or absorption of insufficient quantities of nutrients essential for maintenance and normal growth (i.e. mean liveweight gain ± 2 SD). The diseases and their causes are therefore chronic. Affected cattle grow more slowly than normal, fail to deposit fat, lose condition and weight, become infertile, and, in extreme cases, become emaciated and die.

Reproduction can also be affected by acute and subacute nutrient deficiencies which are induced by a falling food intake and/or by a rising demand for nutrients and metabolites which are too rapid for compensation by the body's homeostatic mechanisms. Such infertility is, by definition (Payne, 1977), a metabolic disease and differs little in principle from the recognized metabolic diseases of cattle, hypocalcaemia, ketosis and hypomagnesaemia. I have adopted the generic term 'metabolic infertility' for the group of acute-nutrient-imbalance-induced causes of infertility (McClure, 1968).

Cattle production is most efficient when cows calve for the first time as soon after they reach the age of two years as is necessary to comply with herd breeding policy, and thenceforth regularly at 12-monthly intervals until they are culled for reasons of low production, surplus stock or old age. In the short term, delay in calving reduces the yield of milk and meat and the profitability of the herd. In the long term, forced culling for infertility and sterility prevents cows from being culled for low production and limits the scope for genetic improvement of the herd.

Efficient cattle production can be achieved by: (i) applying the principles of management for high and efficient production to the individual herd; or, less effectively, by (ii) identifying inefficiency early enough so that the cause can be diagnosed and treated or controlled to prevent significant loss. While the former method is obviously better, managers are not always able to control the herd management and environment sufficiently well and need to back their attempts to do so with the monitoring and problem-solving methods of the latter.

Consultants must be able to measure the fertility of herds, to identify the reasons for infertility and to recommend, where necessary, economical methods of treatment, control and prevention.

Treatment, control and prevention of nutritional and metabolic infertility in herds require an understanding of mating practices, the methods of measurement of herd fertility, reproductive physiology and nutrition.

Mating Practices

Three aspects of mating practice have important bearings upon the reproductive performance of cattle. These are: (i) the duration of the mating season; (ii) the control exerted over mating; and (iii) the mating method.

The duration of the mating season

The mating season, and therefore calving season, may extend throughout the year (non-seasonal, all-year-round or unrestricted mating and calving seasons), or they may be restricted to one or more short periods in the year.

Seasonal mating

A short breeding season for the whole herd is desirable where the herd is kept for production of beef or milk for manufacture and a large proportion of the feed is derived from fresh pasture. The cows are bred to calve at the most suitable time(s) of the year to make the best use of the pasture growth or to take advantage of any seasonal variation in the prices of the products. An example showing the relationship between the feed requirements and supply in seasonal-calving herds is shown in Fig. 1.1.

There are other advantages in having short breeding seasons; for example, farming operations related to breeding and production, such as the supervision of calving cows, can be concentrated. Restriction of the length of the breeding season enforces the control of herd bulls and ensures sexual rest in all the herd at certain times, both important factors in controlling venereal diseases.

These advantages of restricting the breeding periods are so great that when calving of beef cows over longer periods is desirable for reasons of seasonal price structure or unpredictability or unreliability of rainfall, two or more short breeding seasons separated from each other are preferred to unrestricted breeding.

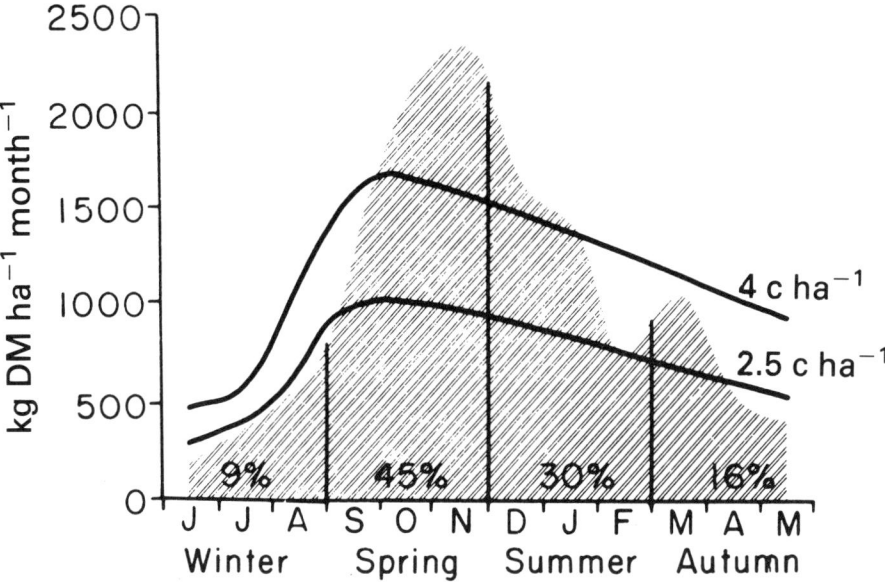

Fig. 1.1. Feed requirements (——) of a winter-calving dairy herd (at two stocking rates, 4 and 2.5 cows ha^{-1}) supplying milk for manufacture in the Waikato district of New Zealand superimposed on the pasture production curve (▨) (seasonal %) (O'Connor, 1982). (Reproduced by permission of the author and the NZ Society of Production.)

Short breeding seasons are successful only if the fertility of the herd is high. The relationship between the pregnancy rate and the duration of the breeding season is exponential. This can be used to determine some important values; for example the number of cows not pregnant at the end of the breeding season can be determined by:

$$A = A_o e^{-kt}$$

where A = the number not pregnant at the end of the breeding season;
A_o = the number of cows being mated;
e = 2.72;
k = log$_e$ pregnancy rate per service;
t = length of the breeding season expressed in terms of the number of possible oestrous cycles.

Thus, with a herd of 100 cows, a pregnancy rate per service of 50% and three services permitted, 12 or 13 cows would still not be pregnant at the end of the breeding season; i.e.

$$A = 100 e^{-\log_e 0.5 \times 3}$$

For ⩾95% of cows to become pregnant during a seven-week breeding season it is necessary for the first- and second-service pregnancy rates to be ⩾79%. If the first- and subsequent-service pregnancy rates are as low as 50%, the breeding season would have to extend over five cycles. When allowance is made for the cows which return to oestrus after cycles of >24 days, the breeding season may have to last for up to five months.

All-year-round or non-seasonal mating

Cows in herds supplying milk for the liquid milk trade must calve at a reasonably uniform rate throughout the year in order to maintain a constant level of production, though a coordinated national milk industry may allow the supply of milk from different areas at different seasons. Strict control over mating is necessary to achieve the planned calving rate and dates. In effect, this means that individual cows are limited to restricted breeding periods.

Assuming a mean cycle length of 21 days, with 100% of cows being mated during the first 21 days after the start of mating and a first-service pregnancy rate of 100%, the mean conception date of the herd would be 10–11 days after the planned first mating date. In order to obtain a mean herd inter-calving (I-C) interval of 365 days and a calving-to-conception interval of 83 days, the cows must be re-bred at the first oestrus on or after 72 or 73 days after calving.

Lower first-service pregnancy rates, which are usual, will give the calving-to-conception patterns as shown in Table 1.1. Also shown are the calving-to-first-service intervals required to produce I-C intervals of 365 days. As the first-service pregnancy rate does not return to normal until about 60 days after calving (VanDemark and Salisbury, 1950), the penalties for low fertility after this time are severe.

Esslemont (1975) concluded it is not possible to achieve the ideal mean 365-day I-C interval (83 days calving to conception) unless there is a 60% pregnancy rate, an 80% heat detection rate and a mean calving-to-first-service interval of <65 days.

Managers of some extensively managed beef herds adopt the practice of allowing bulls to run with cows continuously and at various times of the year cattle are mustered for marking, branding and selecting stock for sale.

The control exerted over mating

Cows may be either paddock mated or hand served. Paddock mating is the term given to the running of the cows with the bull(s) in paddocks without

Table 1.1. Effect of the first-service pregnancy rate on the calving-to-first-service interval necessary to achieve a 365-day inter-calving (I-C) interval.

First-service pregnancy rate (%)	Mean delay in conception (days)	Calving-to-first-service interval (days) required to produce mean I-C interval of 365 days
100	0	73–93
95	1	72–92
90	2	71–91
85	3	70–90
80	4	69–89
75	6	67–87
70	8	65–85
65	10	63–83
60	13	60–80 (a)
55	15	58–78
50	18	55–75
45	22	51–71
40	27	46–66
35	32	41–61
30	39	34–54
25	50	23–43

[a] Fertility is likely to be subnormal with intervals shown below this line.

any form of control being exerted over mating, except for the duration of the mating season. The cows may be run in small groups together with only one bull or they may be run in larger groups with more than one bull.

In dairy herds the maiden heifers are usually run in groups separate from the adult herd, with a bull of their own age or a bull 12 months older than was used on the maiden heifers the previous year. Except when grossly malnourished, these non-lactating maiden heifers are usually fertile.

The disadvantages of paddock mating are that service records are not available, bulls may favour some cows and ignore others, become overworked or injured, or roam, cows may be served that are not wanted pregnant, and selective breeding for genetic purposes becomes difficult requiring small groups of cows and more paddocks.

The advantages of paddock mating are that the cows require less supervision and fewer and less skilful staff. So great are these advantages that paddock mating is practised in many commercial beef herds and in some seasonally mated dairy herds.

With hand service, the cows, when detected in oestrus, are either put in with the bull and service allowed, or artificially inseminated (AI). Hand service does not have the disadvantages of paddock mating, but does require more skill and care in detecting oestrus and does introduce the risk of other causes of reproductive failure, e.g. faulty AI technique.

The rapid increase in the size of dairy herds which is characteristic of the dairy industry today is causing difficulties in the observation of oestrus and hand service/insemination and the recording of data. Such developments have encouraged the use of herd health services and electronic data processing techniques to record and analyse the breeding records.

The mating method

Mating may be by either natural service or by AI. Each method has its own influence upon the reproductive performance and the type of reproductive failure that may ensue.

Natural service introduces the risk of reproductive failure due to male infertility and venereal infectious diseases. Artificial insemination introduces serious risks of infertility resulting from faulty detection of oestrus and from poor insemination techniques.

Standards of Reproductive Performance

Infertility may affect one, a few, or many individuals in a herd. Where only one or a few animals are affected the infertility is described as sporadic. Where sufficient numbers in a herd are affected so as to cause an economic problem, the infertility is described as a 'herd breeding problem' or a 'herd infertility problem'.

Herd infertility can be measured in a number of different ways; often a combination of methods is used for no single method is always sufficient. The methods include measuring:

1. The mean age at puberty and the standard deviation (SD) of the mean.
2. The mean age at first calving and SD.
3. The mean interval between calving and the first post-partum oestrus and SD.
4. The mean interval and SD between successive oestrous periods and the frequency distribution.
5. The number and proportion of cows which do not return to service and therefore are presumed pregnant (the non-return rate) at specified intervals after first service, e.g. 28 and 49 days and at three to four months. These cows are sometimes described as having 'held to service'. Where accurate pregnancy diagnoses can be made (say at 6–12 weeks) the first-service pregnancy rate[1] is preferable to the first-service non-return rate[2].
6. The non-return or pregnancy rate for a specified number of services

---

[1] (No. of cows confirmed pregnant to first service/No. of cows served)×(100/1).
[2] (No. of cows not returning to oestrus after first service/No. of cows served)×(100/1).

or a period of time, e.g. during a three-month breeding season, or by three months after first service.

7. The number and proportion of the cows in a herd that abort.
8. The number and proportion of the cows in a herd that calve prematurely.
9. The number and proportion of calves that die at or about parturition (perinatal mortality rate).
10. The herd mean I-C interval (calving index) and SD.
11. The calving rate.
12. The branding rate.
13. The weaning rate.
14. The culling rate for sterility.
15. The fertility index or service/conception ratio.

The choice of methods will depend upon breeding management: seasonal/non-seasonal, intensive/extensive, hand service/paddock mating. The indices (10) to (14) are of value in determining economic consequences, the others assist in defining infertility problems and diagnosing causes.

Just as individual cows are expected to become pregnant when given the opportunity, so are herds of cows. The expected pregnancy rate, No. of cows pregnant × 100/No. of cows joined[3], is therefore 100% over a period covering one normal oestrous cycle length. Any reduction constitutes some degree of herd infertility for which causes must exist.

A minor degree of herd infertility is generally acceptable but a major reduction in herd fertility is not. The dividing line between what is acceptable and what is unacceptable to the cattle producer is difficult to define. There are at least three approaches that can be made to defining fertility and infertility or sterility in herds of cows.

1. Based on the normal frequency distribution common to many biological systems. The normal range of fertility is taken as the population mean ± 2 SD. Infertility is defined as a fertility level which is $<(\bar{x} - 2\ \text{SD})$.
2. Based on economic criteria. This method has considerable merit, particularly in the extensively managed herds where low reproduction rates can be tolerated because of low production costs. It has the great disadvantage that the level must be determined for each property and possibly for each year.
3. While it is unrealistic to expect a herd fertility rate of 100%, we can expect to be able to raise the level of fertility to the equivalent of the top herds (subject to economic constraints) (Young, 1965). The objective is then to raise the fertility to $>(\bar{x} + 2\ \text{SD})$.

Standards based on economic criteria are shown in Table 1.2.

---

[3] Number of cows running with the bull(s) or the number of cows in the herd to be inseminated or hand served.

Table 1.2. Standards of reproductive performance.

Criteria	Normal	Infertile	Reason for definition[a]
Age at puberty			
Jersey	9 months	> 15 months	(1)
Holstein-Friesian	12 months	> 15 months	(1)
Onset of oestrous cycles after parturition	21–42 days	> 60 days	(2)
First-service non-return rate (49-day basis)	85%	≤ 60%	(3)
First-service pregnancy rate	80%	≤ 50%	(3)
% of herd pregnant by three months after first service	95%	≤ 90%	(3)
Service : pregnancy ratio	1.3 : 1	≥ 2 : 1	(3)
Calving index (herd mean I-C interval)	365 days	≥ 400 days	(4)
Calving rate	95%	< 90%	(5)

[a] (1) conception at 15 months of age is necessary for first parturition at two years of age; (2) mating at first oestrus after 60 days post-partum is necessary for I-C interval of 365 days, if first-service pregnancy rate is 60% (Table 1.1); (3) to achieve a short calving season or to have cows calve close to the planned date; (4) to achieve annual calvings; and (5) economic.

Monitoring Reproduction

The recording and analysis of breeding data is necessary in order to measure reproductive performance, to provide early warning of impending herd infertility problems and to provide a history for the formulation of tentative diagnoses as to the aetiology of the problems should they occur. The data required are listed in Table 1.3.

These data can be recorded manually as shown in Table 1.4 or electronically on a personal computer, using one of the many commercially available programs.

Reproductive failure in individual cows is readily identified in these records and classified. For example, Cow No. 1 in Table 1.4 is a 13-year-old cow which was inseminated less than 60 days post-partum, failed to hold and returned to oestrus after an interval (cycle length) of 18 days.

The breeding records are analysed either manually or by computer and the performance of the herd compared with the objective (Table 1.2). For example in the seasonally mated Herd A (Table 1.4) 93% of cows were

Table 1.3. Data required for monitoring reproduction.

Essential	Desirable
Identification	Origin (home reared or introduced and, if the latter, date of introduction and source)
Age and lactation number	
Lactational state	Yield and composition of milk at the time of mating
Last calving date and disease history at calving, e.g. abortion	
Subsequent oestrous dates	Abnormalities of oestrus; name of observer of oestrus
All mating dates, with identification of bulls or AI and AI details (bull, semen batch, technician)	
Date and results of pregnancy diagnosis	Name of clinician and technique
Abortion/perinatal mortality	

mated within the first four weeks of the mating season, 35% did not return to oestrus after first service and 13% were still not pregnant at the end of the mating season, thus indicating a herd infertility problem. Analyses should be performed regularly at intervals determined by the duration of the mating season, the size of the herd and the intensity of management so as to obtain the earliest possible warning of developing problems.

Intensively managed herds with seasonal calving (e.g. Herd A, Table 1.4)

In an intensively managed, seasonally mated herd the breeding records should be analysed:

1. Just before the beginning of mating, to determine the proportion of cows which have shown oestrus during the preceding four weeks. If it appears that a herd anoestrous or suboestrous problem is present the cause should be determined and control measures applied.
2. At the end of the fourth week after the start of mating, to determine the proportion of the herd that has been mated and to determine the results of mating during the first week of the breeding season, as shown by the 21–24-day non-return rate. Though far too early to give a reliable estimate of the herd fertility, the analysis may give an early indication of a developing herd infertility problem. The cause can then be diagnosed and control measures applied with a minimal delay in conception and subsequent economic loss.

Table 1.4. Herd mating records of a pasture-fed late-winter-calving dairy herd in New Zealand supplying milk for manufacture (Herd A).

Cow number	Age at calving (year)	Last calving date	Calving disease[a]	Pre-mating oestrous dates		Oestrous and mating dates[a]					Pregnancy diagnosis 28 Feb[a]
				Aug	Sept	Sept	Oct	Nov	Dec	Jan	
1	13	17 Aug	RFM				2A 20A				
2	13	20 Jul				28A	10A 19A				
3	12	18 Jul					4A 27A				
4	12	13 Jul				30A					
5	12	6 Jul					3A	9A	18B		
6	12	22 Jul					13A	5A			NP
7	11	2 Jul					4A 24A	17A	11B		NP
8	11	1 Aug					16A				
9	11	19 Aug					10A 29A				
10	11	13 Aug				28A					
11	11	21 Aug					2A 17A	5A 24A	13B		NP
12	10	15 Aug					7A 30A				
13	10	30 Sep					20A	12A			
14	10	23 Jul					3A 18A	2A			
15	10	2 Aug					15A		11B		
16	10	10 Jul				28A			20B	11B	
17	9	19 Jul					7A		18B		
18	9	4 Sep					13A	2A 21A			NP
19	9	8 Jul					12B				
20	9	18 Jul					1A				

#		Date									
21	9	28 Jul		6A	24A						
22	9	27 Jul		4A							
23	8	27 Jul		3A							
24	8	11 Jul		14A							
25	8	17 Aug		2A	25A	6B	15B	29B			
26	8	15 Aug	29A								
27	8	13 Sep		17A		26A	1A	28B	22B		
28	8	20 Jul		7A					2B		
29	7	15 Jul		2B	17B						
30	7	10 Aug		20B							
31	7	3 Aug		13B	29B	16B	15B				
32	7	2 Sep								NP	
33	7	17 Aug		18B		3B				NP	
34	7	15 Aug		16B		1B	21B				
35	7	19 Jul	30B	19B		16B					
36	6	17 Aug		3B	17B	4B		14B			
37	6	4 Aug		15B		4B					
38	6	3 Jul		7B	28B						
39	6	17 Jul		11B		5B		1B			
40	6	20 Jul		5B							
41	6	13 Jul		10B	24B	10B					
42	6	1 Aug		18B		3B				NP	
43	5	20 Aug		4B							
44	5	13 Aug		30B							
45	5	19 Aug		23B		9B	22B				
46	5	3 Jul		14B		2B	20B				
47	5	2 Sep		20B							
48	5	4 Aug		17B		5B			2B		
49	5	13 Aug		13B	29B	13B					2B
50	5	14 Jul		1B	19B	2B					

Table 1.4. (continued).

Cow number	Age at calving (year)	Last calving date	Calving disease[a]	Pre-mating oestrous dates		Oestrous and mating dates[a]					Pregnancy diagnosis 28 Feb[a]
				Aug	Sept	Sept	Oct	Nov	Dec	Jan	
51	4	3 Jul					16B				
52	4	11 Aug					13B				
53	4	14 Aug					21B	10B	20B	6B	NP
54	4	13 Jul					19B	3B 18B			
55	4	25 Aug				29B	17B				
56	4	10 Aug					23B				
57	4	15 Aug					17B		12B		
58	4	3 Aug					15B				
59	4	3 Jul					4A 20B	4B			
60	4	1 Aug					21B				
61	3	26 Jul					8A				
62	3	16 Jul				30A					
63	3	3 Jul									
64	3	13 Jul					4A 27A	12A 23A	14B		NP
65	3	9 Aug					19A	11A			
66	3	10 Jul					2A	9A			
67	3	18 Jul					13A				
68	3	15 Jul					17B	10B 21B			
69	3	28 Aug						3B 20B			NP
70	3	26 Jul					9B 27B		2B		

Cow	Lact.	Calving date					
71	3	19 Jul			23B	19B	12B
72	3	2 Sep			4B 29B	27B	
73	3	20 Aug			26B	9B 24B	
74	2[b]	18 Jul		30B			
75	2	12 Aug		30A	18A		
76	2	11 Jul			15A		
77	2	20 Jul		28A	20A		
78	2	27 Jul			4A		
79	2	17 Jul	DYS		12A		
80	2	18 Jul			15A		
81	2	25 Jul			13A		
82	2	29 Aug			14B	3B 28B	NP
83	2	14 Aug			21B		
84	2	22 Jul	DYS		15B 16B	10B	
85	2	7 Jul			18B		
86	2	4 Aug				3B	
87	2	15 Aug			20B	9B 28B	
88	2	13 Jul		29B	16B 30B	20B	

[a] A = artificial insemination; B = natural service; DYS = dystocia; NP = not pregnant; RFM = retained fetal membranes.
[b] Two-year-old cattle are first lactation heifers (L = 01).

3. At the end of the breeding season, preferably after pregnancy diagnosis six weeks after the last service/insemination, to determine the reproductive performance of the herd and the expected calving distribution for the next season.
4. At the end of the calving season, to determine the incidence of abortions, perinatal deaths and the calving rates.

An analysis of Herd A is shown in Table 5.2, p. 68.

Intensively managed herds with all-year-round calving

The breeding records of intensively managed herds with all-year-round calving should be analysed at regular monthly, or more frequent, intervals.

At each analysis, the following should be determined:

1. The proportion of the herd that has/has not been observed in oestrus by 60 days after parturition, and/or the mean interval (and SD) between calving and first oestrus.
2. The mean interval between calving and first service (and SD) and/or the proportion mated/not mated by 83 days after parturition.
3. The first-service pregnancy/non-return rate.
4. The mean parturition-to-conception interval (and SD) of those cows confirmed pregnant.
5. The proportion not pregnant by three months after first service, and the proportion culled for reproductive failure.
6. Calving data (including abortion and perinatal mortality rates).

An example of the breeding records of an infertile herd (Herd B) with an all-year-round calving season is shown in Table 5.3, p. 74, and the analysis in Table 5.4, p. 79. There are both advantages and disadvantages in working with herds with extended calving and mating seasons. Chronological analysis enables seasonal trends in fertility to be detected. If the seasonal trend in fertility can be matched to seasonal changes in feeding (quality or quantity) and/or production, a tentative diagnosis of nutritional and metabolic infertility may be possible. The main disadvantage is that the spreading of the breeding reduces the number of cows available for comparison each month.

Herd B had a chronic infertility problem extending over 1981 and 1982, characterized by an apparent delay in the recurrence of oestrous cycles after calving, delayed mating, a low first-service pregnancy rate and delay in the establishment of pregnancy. These resulted in a long mean inter-calving interval. Over 10% of cows retained the placenta after calving and cows returned to oestrus after an unsuccessful first service after abnormally long periods.

Extensively managed herds

Monitoring of the reproductive performance of extensively managed herds is based solely on the retrospective analysis of breeding records, and yields little more than the mean I-C interval, the calving rate and the weaning rate.

This is sufficient to identify herd infertility problems but not in time to diagnose and treat/control incipient problems. The only satisfactory approach is to seek additional data before and during the mating period. This is considered again in Chapter 6.

Reproductive Physiology

The physiology of reproduction relevant to the subject of nutritional and metabolic infertility in the cow is summarized here. Further information is available in the reviews by Short and Adams (1988), Butler and Smith (1989), Ferguson and Chalupa (1989), Randel (1990), and in the supplements to the *Journal of Reproduction and Fertility*, numbers 30 (1981), 34 (1987) and 43 (1991).

During pregnancy, oestrogen and progesterone are secreted by the placenta and circulate in the maternal blood, inhibiting the secretion of gonadotrophin-releasing hormone (GnRH) by the hypothalamus.

The gonadotroph cells of the adenohypophysis are then deprived of sufficient stimulation to maintain the synthesis of follicle-stimulating hormone (FSH) and luteinizing hormone (LH). These must be restored after parturition before normal cycles can commence (Nett, 1987). Restoration is completed by day 10 (D10) (Moss *et al.*, 1985), and episodic release of LH normally commences on D13 (Wagner and Hansel, 1969; Peters *et al.*, 1981).

Episodic release reaches a peak mean concentration and frequency two weeks before the first post-partum oestrus which occurs between D17 and D42 (Humphrey *et al.*, 1983).

Oestrous cycles average 20-21 days in length with 85% ranging from 18 to 24 days (Asdell, 1964). Each is divided into a luteal phase of 17 days and follicular phase of three to four days.

Behavioural oestrus, characterized by receptivity to the male, homosexual and other behavioural changes, and the discharge of oestrous mucus from the vagina, lasts for 15-18 hours (range 12-26 hours) with ovulation occurring 10-12 hours after the end of oestrus or 25-30 hours after the pre-ovulatory surge of gonadotrophin.

The fertilized ovum cleaves daily after fertilization, reaches the uterus at 72-84 hours in an 8-16-cell stage of development, and commences to attach to the endometrium at 22 days with interdigitation between fetal and maternal tissues by D30 (King et al., 1979). Pregnancy lasts 282 ± 5.6 days (with some breed variation).

The sequence of events causing these gross signs of reproductive activity in the cyclic cow can be said to commence in late dioestrus (or the late luteal phase of the cycle). At this time the luteal cells secrete oxytocin which stimulates the endometrium to produce pulses of prostaglandin $PGF_{2\alpha}$ (Bazer et al., 1991) which inhibit the LH activation of adenyl cyclase by the luteal cells causing them to stop secreting progesterone. This removes the inhibitory action of progesterone on the hypothalamus and adenohypophysis and allows the GnRH, which is being released in small amounts at approximately hourly intervals from the arcuate nucleus in the hypothalamus (episodic or tonic release) (Halász and Pupp, 1965; Dyer, 1985), to stimulate the release of FSH and LH from the gonadotroph cells of the adenohypophysis.

FSH binds on to receptors in the thecal and granulosa cells in the ovary, stimulating mitosis of the granulosa cells and follicular growth, and the synthesis of pregnenolone from cholesterol in the thecal cells, then oestradiol-17β from the pregnenolone in the granulosa cells.

Oestradiol-17β sensitizes the hypothalamus to further increments of oestradiol-17β and the gonadotroph cells of the adenohypophysis to GnRH. It also further stimulates the granulosa cells of the ovary and the secretion of more oestradiol-17β. This then acts on the preoptico suprachiasmatic anterior hypothalamic tubero infundibular system (Saito, 1981; Barraclough and Wise, 1982) causing release of a burst of GnRH into the portal vessels at the median eminence ('pulsatile release') (Cross and Dyer, 1972; Drouva et al., 1984).

The pulsatile release of GnRH stimulates the release of a large quantity of LH ('LH surge') by the gonadotroph cells of the adenohypophysis which have been sensitized by oestradiol-17β (Cooper et al., 1974; Gordon and Reichlin, 1974).

During this surge, the LH binds on to receptors on the thecal and granulosa cells which had developed in response to FSH stimulation and stimulates: (i) the further development of follicles; (ii) the resumption of meiosis and the pre-ovulatory maturation of the oocyte; (iii) the shedding of the secondary oocyte with chromosomes arranged at the second metaphase; (iv) extrusion of the polar body; and (v) ovulation and the conversion of the granulosa cells to lutein cells and secretion of progesterone (Donaldson and Hansel, 1964). GnRH appears to have some direct effect on the maturation of the ovum (Erickson et al., 1983).

After fertilization, the trophoblast secretes trophoblast protein-1 (TP-1) from D15 (when the spherical trophoblast begins to elongate), reaching its maximum rate on D16 to D19 and continuing until at least D38. TP-1 appears to bind to receptors on the endometrial cells inhibiting the pulsatile release of $PGF_{2\alpha}$, thus inhibiting luteolysis (Helmer et al., 1989a, b; Bazer et al., 1991; Roberts et al., 1991).

The function of the reproductive endocrine system is modulated by nervous signals from other hypothalamic nuclei and higher centres, the ovarian hormones oestradiol, progesterone and inhibin and the endogenous opioid peptides (Meites, 1984; McQueen and Fink, 1988; Sprangers and Piacsek, 1988).

In the 21-day cyclic cow the concentration of progesterone falls from its peak of 6–8 ng ml^{-1} to <1 ng ml^{-1} 16–19 days after oestrus. The plasma oestradiol-17β concentration rises from its minimum of <10 pg ml^{-1} (mean 3.6 pg ml^{-1}) to a peak of 15–25 pg ml^{-1} on the day before next oestrus and falls within two to five hours after the beginning of oestrus to its basal level. The concentration of LH in the plasma rapidly increases from its basal level of 2–3 ng ml^{-1} to >10–15 ng ml^{-1} for six to eight hours with a peak of 10–65 ng ml^{-1} corresponding to the onset of oestrus. Progesterone appears in the blood in appreciable quantities five days after oestrus and continues to increase until about the 16th or 17th day by which time it has reached its peak of 6–8 ng ml^{-1} (Stabenfeldt et al., 1969).

The metabolism of the cells of the reproductive organs, ovum, embryo and fetus appears to be similar to those of other systems. Though of fundamental importance to the understanding of nutritional and metabolic infertility, the details are outside the scope of this book except that attention needs to be drawn to the critical role of energy in the metabolism of the reproductive system.

This is because of the demands for large amounts of energy for fetal development in the last trimester of pregnancy and for milk secretion. Fetal development and lactation also drain amino acids and minerals from the maternal circulation. At times these may also become critical.

Energy for cellular metabolism is derived from adenosine triphosphate (ATP) which is replenished via the terminal oxidative pathways in the tricarboxylic acid (TCA) cycle from glucose and the free fatty acids (FFA). Availability of oxaloacetate (derived from glucose) appears to be the rate-limiting factor for the incorporation of FFA into the TCA cycle. Glucose and the FFA themselves are derived from propionate, lactate, acetate and amino acids which have been absorbed from the alimentary tract.

The conversion of these substrates to the glucose and FFA which are metabolized, involves many enzymes, coenzymes and cofactors, parts of which are essential dietary nutrients. This is described in Chapter 2.

The cells of all reproductive organs, ovum, embryo and fetus metabolize glucose or the products of glycolysis for the replenishment of ATP (Brinster, 1967, 1970; Whittingham and Biggers, 1967; Setchell et al., 1972; Biggers and Borland, 1976; Sen et al., 1979; Liggins, 1982; McLaren, 1982; Mazur and YoungLai, 1986; Fagbohun and Downs, 1992). Insulin is required to assist the entry of glucose into all cells other than nervous, e.g. luteal cells require insulin for maximal progesterone production (Poff et al., 1988).

The hypothalamus, ovum, embryo and fetus have an obligatory

requirement for glucose, and as late-pregnant and lactating cows have difficulty in maintaining glucose homeostasis any reduction in availability of glucose is likely to affect their metabolism, and possibly the survival of the ovum and embryo.

Acetate, the FFA and amino acids may have a significant, but as yet undefined, role as an alternative energy source for the adenohypophysis and the ovary (Teleni et al., 1984, 1985, 1989; Downing and Scaramuzzi, 1991).

Sodium, potassium, calcium and magnesium ions have critical roles in the metabolism of the cells of the reproductive system but they have not as yet been shown to be involved in nutritional and metabolic infertility. Their involvement in other metabolic diseases (Payne, 1977) suggests that they may also be capable of affecting reproduction.

The aspects of reproductive physiology which are relevant to this discussion on nutritional and metabolic infertility include:

1. The re-establishment of the ovarian cycles post-partum. Episodic release of LH, and consequently the first post-partum oestrus can be delayed by a negative energy balance, lactation and suckling (Butler and Smith, 1989). A short luteal phase is common after the first post-partum oestrus (Perry et al., 1991).
2. Inhibition of pulsatile release of LH by acute energy deficiency (McClure and Saunders, 1985; Randel, 1990).
3. The intensity of oestrus. The intensity and duration of oestrus is of critical importance to successful AI in cattle. Field and experimental evidence suggests that nutrition can affect both intensity and duration (Suzuki et al., 1982).
4. Ovarian function. In nutritionally restricted beef cattle follicular development can be arrested at the medium-sized stage (Rutter and Manns, 1991).
5. Fertilization, development of ovum, embryo and fetus. Up to 35–40% of ova and embryos die before implantation and mostly before D16 (Ayalon, 1978; Diskin and Sreenan, 1980; Humblot et al., 1988) for reasons which are inadequately understood but which appear to include undernutrition (Hill et al., 1970) and may involve hypoglycaemia (Weigelt et al., 1988). Death of the embryo or ova after D16 extends the luteal lifespan and delays the return of oestrus. Cows in which fetal death occurs between D24 and D50 return to oestrus a fortnight later (13.5 ± 0.8 days, Ryan et al., 1992).

2

CATTLE NUTRITION

Though primarily grazing animals, cattle are able to eat and digest feedstuffs of widely ranging type and composition. These vary from those suitable for monogastric animals to coarse fibrous fodder requiring the aid of bacterial fermentation to break down the plant cell-walls. The essential role of cattle in human society is to convert these fibrous plants, together with certain industrial by-products, into protein for human consumption. Commonly, however, land suitable for the cultivation of crops for direct human consumption is used for growing crops for cattle feed; a practice driven by consumer demand.

Beef and dairy cows have relatively modest food requirements during growth and are not likely to be affected by malnutrition except when they are fed on the following: browned-off mature plants after seed-fall; insufficient amounts of feed; plants which have been grown on soils containing insufficient available essential trace minerals not required for plants; or plants which contain toxic substances.

Cows that are pregnant, particularly whilst still growing, and those that are lactating are vulnerable to deficiencies of those nutrients which are required in large quantities by the fetus during the last two months of gestation and by the mammary glands for milk synthesis.

The difficulties which ruminants have in balancing intake with requirements have been aggravated by the selection of very high-producing cows which can milk at levels beyond their digestive capacity, causing the cows to draw excessively on their body reserves during the early part of lactation before their intake has reached its potential.

Though it would seem, at first glance, that the quantity of digestible organic matter ingested would be a sufficient guide as to the adequacy of the diet, this is not always so. Some nutrients essential for the cow are not required by the ruminal flora or else are required in different proportions. There is a variable loss of nitrogen (N) in eructated ruminal gas. Also, the composition of the dietary carbohydrate and protein affects the cow's energy metabolism.

After weaning, cattle are fed almost exclusively upon plants and plant products. Some are fed on the vegetative parts of plants, others on the seeds and still others, perhaps the majority, on combinations of these, sometimes with the addition of animal products. Cattle fed exclusively on the vegetative parts are subjected to diets which vary both in quantity and quality and which at times do not meet their nutritional requirements. These variations are often outside the immediate control of the farm manager.

On the other hand, and with one major exception (the very high-producing cow), feeds manufactured from plants and animal by-products can be formulated and fed in quantities which meet the animals' requirements at their various stages of growth, reproduction and levels of milk production. The composition of these feedstuffs and the amounts fed are within the control of the manager.

Feeding Practices

There are four basic feeding practices.

Pastures/crops

Where both land and suitable climate are available, pasture grown under 'dry land' conditions forms the cheapest foodstuff for cattle and therefore provides the food for beef-breeding herds and dairy herds supplying milk primarily for manufacture. The pastures may be grazed by the cattle, cut and fed in feeders ('zero-grazed') or cut and conserved for feeding out later when pasture growth is temporarily insufficient. Where perennial pasture is not sufficiently productive under 'dry land' conditions, it may be supplemented with or replaced by crops or irrigated pastures and crops which, though more expensive, are usually cheaper than manufactured feedstuffs. In some beef-breeding areas, e.g. those with semiarid climates, herbs, shrubs and even the leaves of trees may form part or all of the diet.

Compounded feedstuffs

Some dairy cows supplying milk for the liquid milk trade are kept close to markets provided by cities and towns in districts which are unsuitable for pasture production either under 'dry land' or irrigation conditions. Feedstuffs for these cattle consist of balanced concentrate feeds made from cereal grains and protein-rich seeds, by-products, vitamins and minerals, together with minimal quantities of purchased roughage, formulated by simple arithmetical means, or more conveniently using computer programs such as 'CAMDAIRY' (Hulme et al., 1986).

Combination of concentrates and pastures/crops

Probably the greatest proportion of dairy cows supplying milk for the liquid milk trade are fed on pasture or pasture products and crops to meet their maintenance and some productive requirements and are fed concentrates for the remainder. This trend towards feeding cattle on manufactured feedstuffs is intensified in those parts of the world where land suitable for pasture production is limited or the climate is unsuitable for lengthy periods of pasture growth.

By-products

An increasing number of cattle are fed on by-products and wastes of the plants used in the fruit and vegetable canning and freezing, and beverage and fibre industries (Barber and Lonsdale, 1980; BSAP, 1980).

Feed Intake

The intakes of digestible organic matter (DOM) and essential minerals are major factors affecting production. The intake of DOM depends upon appetite, availability and the chemical composition of the feed.

Appetite

Appetite is a function of the animal's inherent craving for food, and the sum of stimuli arising from stretch receptors in the alimentary tract, chemoreceptors sensitive to pH and acetate concentration in the ruminal wall and sensitive to propionate in the liver, transmitted to the brain via the vagal and splanchnic nerves (Forbes, 1986). Acetate, butyrate and propionate, but not glucose, act as signal compounds determining appetite, and the rumen fill at any one time is a function of appetite, available feed and digestibility. The magnitude of the appetite depends upon palatability (Raymond, 1964) and the animal's nutrient requirements, which are determined by genotype, age, liveweight, body composition, climate, physiological state and milk yield (Greenhalgh and Runcie, 1962; Hutton, 1962, 1963; Campling, 1964; Minson, 1982; Weston, 1982; Treacher et al., 1986).

Ruminants adjust their voluntary food intake, i.e. the amount of food eaten when they are offered food *ad libitum*, according to physiological demands for energy until limited by fill or rumen load (Campling, 1964;

Montgomery and Baumgardt, 1965a,b; Simkins et al., 1965a,b; Bines and Davey, 1970). The limitations of abdominal space and therefore rumen capacity, imposed by the developing fetus, membranes and fluids in late pregnancy, restrict intake, have a continuing effect after parturition and are still present during the normal breeding period (Fig. 2.1).

Appetite and intake of dry matter (DM) are reduced when cattle and sheep are offered diets which contain >86% intracellular water (Davies, 1962) or insufficient of one or more essential nutrients, are denied access to sufficient drinking water, are exposed to extreme climates (with ambient temperatures <10°C and >30°C, particularly when changeable), are diseased, e.g. with intestinal parasitism, and when travelling time to and from water is excessive (Forbes, 1986).

While the upper limit to intake is set by appetite, the actual intake is determined by availability and chemical composition of the feedstuff.

Availability

Availability of feed for grazing cows is a function of stocking rate, pasture area, length and density, and duration and frequency of grazing. For zero-grazed and concentrate-fed (hand-fed) cows it is a function of quantity offered, duration of access, frequency of feeding and trough space. The number of feeding periods each day needed and the maximal permissible intervals between feeding depend upon the composition of the feedstuffs. The frequency of feeding high-grain diets influences blood propionate, glucose, insulin and growth hormone concentrations (Sutton et al., 1988).

Chemical composition

The rate at which food is digested in the rumen and passes on to the small intestine is an important determinant of intake (Blaxter et al., 1961; Corbett et al., 1963; Nelson et al., 1968; McClure, 1977a; Hodgson, 1982; Minson, 1982, 1990). This digestibility is influenced mainly by the extent to which the fibre in the plant cell-walls has been lignified (Wilson et al., 1966) and the activity of the ruminal microflora which, in turn, depends upon their having adequate readily available carbohydrate, non-protein N or rumen-degradable protein and essential minerals in the diet.

Feed Composition

Tables showing the average composition of the many feedstuffs fed to cattle have been published for most countries with developed cattle industries. Examples include the USA (NRC, 1982), the UK (McDonald et al., 1991; MAFF, 1992) and Australia (Ostrowski-Meissner, 1990).

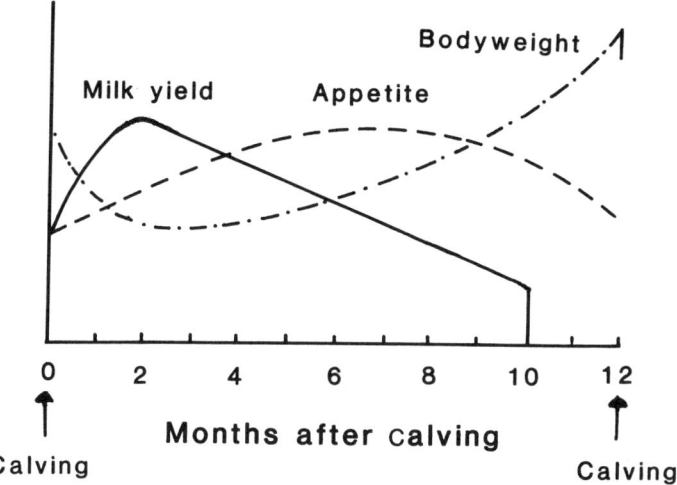

Fig. 2.1. A general relationship between milk yield, bodyweight and appetite of dairy cows after calving (Haresign, 1979). The interval between parturition and peak intake is influenced by food composition. (Reproduced from *Recent Advances in Animal Nutrition* Copyright 1979, by permission of Butterworth and Company (Publishers) Ltd, and by courtesy of Dr W. Haresign.)

While the data supplied in these tables provide a useful guide to the composition of feedstuffs and the planning of rations for cattle, it is necessary to take into account the fact that the tables show average values, not always the range. Neither do they indicate the nutritive value as distinct from its chemical composition. Nutritive value or 'quality' of a feedstuff is the chemical composition relative to the requirements of the class of cattle under consideration. These requirements vary according to the various animal factors described under Feed Intake above.

Pasture and forage crops

The most important factor affecting the composition of pasture grasses and forage crops is the stage of growth or maturity (Smith, 1960; Weinmann, 1961; Waite *et al.*, 1964; Wilson and McCarrick, 1967; Weston and Hogan, 1968a,b; McClure, 1970). The change which occurs during the growth and maturation of plants is most marked in the annual species and is best illustrated by referring to annual crops such as forage oats (Figs 2.2 and 2.3) and annual ryegrass (Table 2.1). The most nutritious stage of growth occurs at the immediate pre-flowering stage when the digestible carbohydrate, protein and minerals are at or close to their peak concentrations and digestibility is still high. After flowering the bulk of the nutrients is

Table 2.1. The chemical composition (% DM) of ryegrass cut at differing stages of growth.[a]

	Cut 1		Cut 2		Cut 3		Cut 4	
Ether soluble		9.1		7.6		6.5		4.7
Hexoses	4.6 ⎫		6.3 ⎫		6.7 ⎫		4.0 ⎫	
Sucrose	3.2 ⎬	13.8	2.7 ⎬	11.8	2.8 ⎬	11.3	2.8 ⎬	10.6
Fructosan	6.0 ⎭		2.8 ⎭		1.8 ⎭		3.8 ⎭	
Organic acids		4.2		4.9		4.6		2.9
Protein N×6.25	14.6 ⎫	18.5	11.6 ⎫	15.2	10.6 ⎫	13.8	6.4 ⎫	9.6
NPN×6.25	3.9 ⎭		3.6 ⎭		3.2 ⎭		3.2 ⎭	
Cellulose		21.3		22.1		23.9		26.7
Hemicelluloses comprising:								
xylan	6.9 ⎫		8.9 ⎫		10.1 ⎫		14.4 ⎫	
araban	2.1 ⎪		2.2 ⎪		2.3 ⎪		2.5 ⎪	
glucan	1.9 ⎬	15.8	2.2 ⎬	18.9	2.0 ⎬	19.4	1.7 ⎬	25.7
galactan	0.7 ⎪		0.7 ⎪		0.7 ⎪		0.7 ⎪	
aldobiouronics	4.2 ⎭		4.9 ⎭		4.3 ⎭		6.4 ⎭	
Pectin		2.4		2.1		2.2		2.2
Lignin		2.7		3.6		4.3		7.3
Ash, SiO$_2$-free	6.9 ⎫	8.1	7.6 ⎫	8.5	7.0 ⎫	7.8	5.1 ⎫	5.7
SiO$_2$	1.2 ⎭		0.9 ⎭		0.8 ⎭		0.7 ⎭	
Total		95.9		94.7		93.8		95.4

[a] Ryegrass S23: Cut 1 – young leafy stage of growth; Cut 2 – late leafy; Cut 3 – head emergence; Cut 4 – seed setting. (From Waite et al. (1964), reproduced by permission of The Cambridge University Press.)

concentrated in the seeds which form the basis of manufactured concentrate feedstuffs.

The concentration of readily available carbohydrate decreases during the formation of new shoots, more particularly during spring, and increases towards maturation (Weinmann, 1961). A considerable portion of the total available carbohydrate may be lost due to respiration both during the formation of new tissue and during periods of dormancy, e.g. winter in humid temperate climates, overcast weather, and during wilting, drying and storage.

Other factors which affect the chemical composition of pastures and crops include:

1. Botanical composition (Butler and Johnston, 1957; Butler and Peterson, 1961; Bailey, 1962; Johns, 1962; Ulyatt, 1969; Minson and McLeod, 1970; Miltimore et al., 1975; Hacker and Minson, 1981).
2. Soil structure, texture and chemical composition, and pH.
3. Fertilizer treatment (Bryant and Ulyatt, 1965; Hogan and Weston, 1969; Fernando and Carter, 1970). Nitrogenous fertilizers have little effect on the cell-wall constituents but do decrease the concentration of soluble carbo-

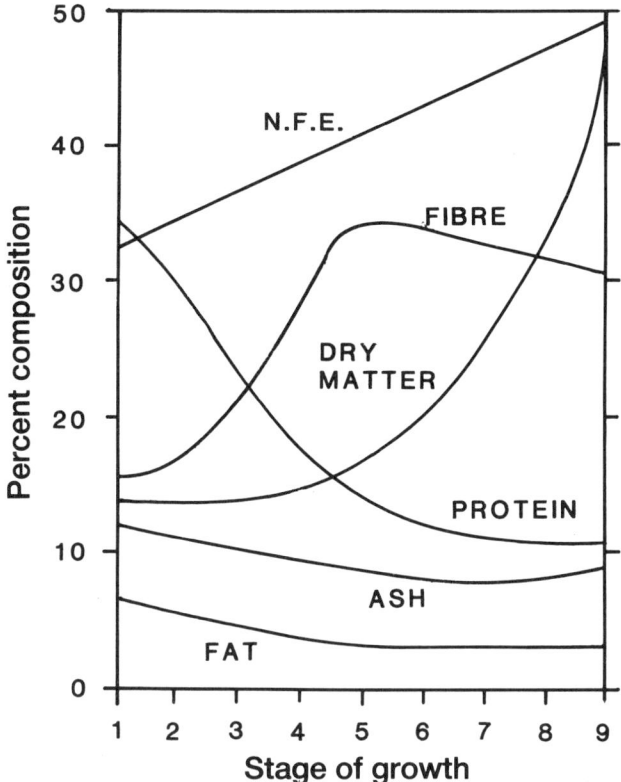

Fig. 2.2. Composition of Clintland oats with advancing maturity (Smith, 1960). Stages of growth and height are as follows: 1 = 4 leaf, 150 mm; 2 = 4 leaf, 275 mm; 3 = 4–5 leaf, 400 mm; 4 = boot-terminal florets visible, 650 mm; 5 = heads emerging, 975 mm; 6 = early milk, 1075 mm; 7 = early dough, 1100 mm; 8 = seed nearly ripe, 1100 mm; and 9 = ripe, 1100 mm. NFE = nitrogen-free extract. (Reproduced from the *Agronomy Journal* (1960) 52, 637–639 by permission of the American Society of Agronomy.)

hydrate whilst increasing the nitrate and total N content (Bryant and Ulyatt, 1965; Hogan and Weston, 1969) and at higher rates can profoundly affect nutritive value.
4. Season (Hutton, 1961; Corbett *et al.*, 1963; Hutton *et al.*, 1967; Wilson and McCarrick, 1967).
5. Rainfall.
6. Ambient temperature (Hacker and Minson, 1981).
7. Spoilage by rain, sunlight, fungi and slow oxidation.
8. Sunlight affecting the soluble carbohydrate and NO_3 content of growing grasses and crops.

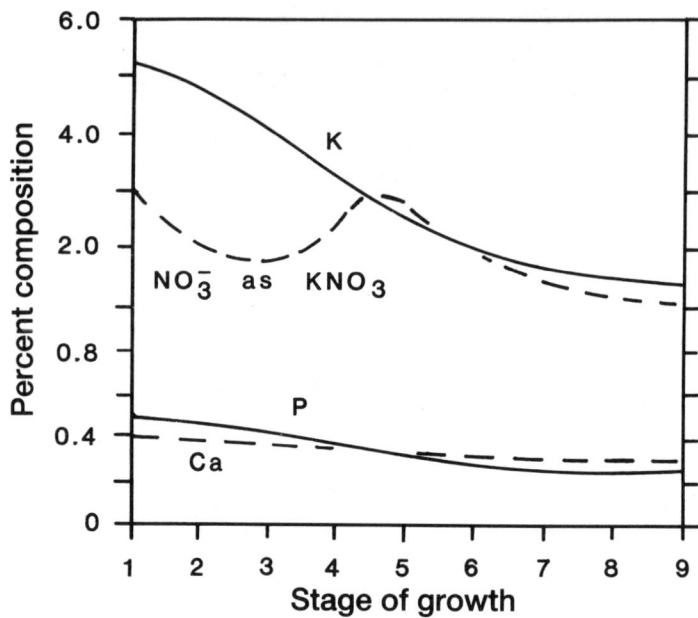

Fig. 2.3. Mineral composition of Clintland oats with advancing maturity (Smith, 1960). (For stages of growth, see Fig. 2.2.) (Reproduced from the *Agronomy Journal* (1960) 52, 637–639 by permission of the American Society of Agronomy.)

All affect digestibility (Hacker and Minson, 1981) and thereby appetite and intake of DOM.

Concentrates

The chemical composition of manufactured concentrate feedstuffs depends upon the composition of the raw materials and their formulation and compounding.

The raw materials are mainly grain, grain by-products and protein-rich by-products of the food and fibre industries. They differ from pasture and forage crops in that most contain little fibre or water, their sources of energy are mainly starch which is readily fermented and their protein is usually less readily degraded in the rumen.

Feed Type

Pasture

Unlike manufactured feedstuffs which can be formulated to meet the requirements of cattle at their various stages of growth, reproduction and lactation and fed at the required levels, pastures and crops are feedstuffs which vary considerably in both quantity and composition. Further, it is not easy for the manager to estimate accurately either the amounts of available pasture or its quality. Deficiencies of both can place the cattle under considerable degrees of nutritional stress which, at the critical stages of reproduction, can cause infertility.

The most productive pastures are those composed of European grasses and clovers, particularly ryegrass and white clover, grown in temperate climates with a high and uniformly distributed rainfall of between 1000 and 1200 mm per annum (or alternatively under irrigation). Such pastures, when properly managed have produced annually 13,500 kg DM ha^{-1} of highly nutritious feed and a ten-year average production of 648 kg milk fat ha^{-1} (Campbell *et al.*, 1977; Campbell, 1982), equivalent to a 90% utilization of the pasture. Jersey cows grazing such pasture have recorded a mean intake of 3.2 kg DM 100 kg^{-1} liveweight and 152 MJ energy day^{-1} and a yield of 19.3 kg of 4% fat-corrected milk (FCM) and 0.88 kg fat over the peak of lactation (Bryant and Trigg, 1982).

Pasture deteriorates if it is not fed to livestock immediately it has reached its optimal stage of growth, and as the stocking rate is the most important factor affecting the conversion of pasture to animal product (McMeekan, 1956), a fine balance in stocking rate is essential if the objective of a maximal yield of protein per hectare is to be achieved without affecting the health of the cattle. This imposes considerable risks of managerial error causing overstocking, food shortage and/or feeding of immature feed on one hand and wasting feed and/or feeding overmature, poor quality feed on the other.

Even when the rainfall (or irrigation) is optimal, the rate of pasture growth varies according to the season because of changes in soil temperature and the number of hours of daylight. Fine control needs to be exerted over the stocking rate so that surplus pasture, usually in the springtime, can be conserved and fed to the cows during periods of shortage, with a minimum of waste. Where the length of the pasture-growing season is unduly limited by severe and long winters or by regular summer/winter droughts it may be necessary to supplement the pasture with crops.

The least productive pastures are those which grow in the tropical and subtropical areas and are composed of coarse species such as *Themeda* spp. and *Heteropogon* spp. These grow rapidly during the wet seasons that occur in these areas but mature and remain as standing 'straw' until the next wet season and tend to be deficient in protein, minerals (particularly phosphorus) and readily available carbohydrate, and contain excessive proportions of crude fibre. At that stage the grasses are indigestible and little is eaten. For example, Robinson and Sageman (1967) reported crude protein falling from 5.0–6.9% in the wet season to 1.3–3.9% at the end of the dry season and apparent digestible energy values and voluntary intakes in sheep as low as 6.7–9.6 MJ kg^{-1} and 11.5–17.2 g kg^{-1} liveweight day^{-1} respectively.

The annual DM yield of such tropical and subtropical pastures can be considerable but their digestibility is low, and although they have produced up to 180 kg liveweight increase and 2500 kg milk ha^{-1} year^{-1} (Mannetje, 1982) much lower yields are usual. However, with irrigation and the application of fertilizers, particularly N, selected tropical species can be very productive (Hutton, 1974).

The best quality pasture, containing 11.8 MJ metabolizable energy (ME) kg^{-1} DM, is capable of supplying sufficient energy for maintenance plus 18–22 kg milk (Holstein-Friesian of 500 kg liveweight) or 22–23 kg FCM when zero-grazed (Meijs, 1981) at the time for breeding, about two to three months post-partum. This milk yield is well below the potential of Holstein-Friesian cattle and is often below the actual yield even in unsupplemented cows. Managers need to ensure that there is no restriction on the quantity of pasture available to the cattle, or on grazing (feeding) time, and that the quality of the pasture is always at the optimum necessary for the stage of growth, production and reproduction of the cows.

Hay and silage

The chemical composition of hay and silage is related to the pasture or crop from which they were made, and though some losses occur in processing, the stage of growth or maturity at the time of cutting remains the main determinant of composition. Some soluble carbohydrate, protein and beta-carotene are lost, and the water content reduced. Further losses occur during storage.

Cattle eat up to 20% less silage DM than hay made from similar herbage (Campling, 1964; Rogers *et al.*, 1979) unless it is very well preserved.

Crops

Crops are grown for cattle fodder in areas in which the climate is unsuitable for permanent pastures and to fill gaps in the year when permanent pastures are not sufficiently productive. Most crops are annuals and as most plants in the crop are at similar stages of growth and maturity when eaten or conserved, the stage of maturity at the time of feed becomes a critical factor in the determination of feed composition and nutritional value. In addition, the effect of any toxic substances or nutritional deficiencies in the crop are fully manifest as they are not diluted by plants of other species or stages of maturity.

The crops commonly grown for cattle are:

1. The cereals, e.g. forage oats, millet, sorghum.
2. Cruciferous feeds, e.g. mangolds, sugarbeet, swede-turnips, white turnips and rape, kale and chou moellier.
3. Lucerne (alfalfa).

By-products

By-products and wastes used in cattle feeding include:

1. Cellulose wastes (Greenhalgh, 1980): straw, sawdust and other wood products.
2. Fruit, vegetable and arable crop by-products (Francis, 1980; Klopfenstein, 1983): citrus pulp, sugarbeet tops, potatoes, Brussels sprouts waste, cauliflower, cabbage, carrots and green pea haulm.
3. Animal excreta (Wilkinson, 1980): poultry, pig and cattle.
4. Cereal, sugarbeet and potato processing waste (Barber and Lonsdale, 1980; Klopfenstein, 1983): brewers' grains, bran, rice, soyabean, cottonseed hulls, maize gluten meal, fermented yeast, beet pulp and molasses.
5. Food and dairy waste (Singer, 1980); oilseed, sugar, fruit and vegetables.
6. Slaughter waste (Cooke and Pugh, 1980; Klopfenstein, 1983): blood, feathers, meat, bone and animal fat.
7. Fish by-products and waste (Pike and Tatterson, 1980; Klopfenstein, 1983).
8. Tropical by-products (Morgan and Trinder, 1980): olive pulp, grape pulp, coffee residues, cocoa residues, citrus pulp and cassava meal.

Manufactured feedstuffs

Feedstuffs are manufactured for cattle from grain and by-products of various industries supplying food and clothing for man. The by-products (peanut meal, soyabean meal, linseed meal, safflower meal, cottonseed meal, rapeseed meal and meatmeal) are protein-rich feeds. Energy is provided by grains grown in the main specifically for animal feed, or inferior grades of grain grown for human feed, e.g. oats, barley, wheat, sorghum. Fibre and some digestible carbohydrates, proteins and minerals are provided by stems and leaves of cereals, grasses and legumes such as lucerne. Minerals and some vitamins are usually also added to complete the balance of nutrients.

Provided that moderation is exercised in the level of pre-partum feeding, it is possible to formulate and compound rations to meet the requirements of the cattle for the various stages of growth, production, reproduction and lactation, and to feed these rations at the necessary levels to ensure that the animals do not develop negative nutrient balances, and do not develop nutritional and metabolic infertility.

However, should potentially very high-yielding dairy cows be overfed ('steamed-up') before calving, and calve in fat condition (score above 3.0/5), it is unlikely that diets containing sufficiently high concentrations of ME can be formulated to supply sufficient energy post-partum to prevent a negative energy balance developing.

Energy

Cows derive their energy mainly from the products of the ruminal fermentation of plant carbohydrates and, to a lesser extent, from fermented and unfermented plant proteins. Fermentation yields acetates, butyrates and propionates. In addition, small quantities of starch and soluble carbohydrates pass through the forestomachs unfermented, to be hydrolysed in and absorbed from the small intestine (Symonds and Baird, 1975).

The nutritive value of the dietary sources of energy depends upon (i) the total quantity of volatile fatty acids produced by the ruminal biota from the ingested carbohydrate and protein; and (ii) the length of the carbon chains or the number of carbon atoms in the fatty acids. In the course of fermentation, mono- and oligosaccharides, fructosans and starch are released rapidly and produce more propionate whilst the celluloses and hemicelluloses (cell-wall components) are released and fermented slowly and yield more of the acetate present in the rumen liquor (Bailey, 1962; Bath and Rook, 1963; Storry and Rook, 1966; Ulyatt, 1969).

Propionate is converted via the gluconeogenic pathways into glucose which becomes available as a substrate for the energy metabolism of all tissues and organs.

In addition to propionate, the glucogenic amino acids derived from unfermented ingested protein ('escape', 'by-pass' or 'undegraded' protein) and from ruminal biota or from catabolized tissue protein, and glycerol derived from catabolized fat are also available for gluconeogenesis (Leng, 1970).

Other nutrients, ingested as such or produced by the ruminal biota from dietary substrates, act as, or are incorporated in, the enzymes, coenzymes or cofactors which are involved in gluconeogenesis, glycolysis, the tricarboxylic acid (TCA) cycle and the terminal oxidative pathways to provide energy for the formation of ATP from ADP. These nutrients are the minerals cobalt, copper, iron, magnesium, manganese, phosphorus, and potassium (Underwood, 1977, 1981) and the vitamins biotin, cyanocobalamin, nicotinamide, pantothenic acid, pyridoxine, riboflavin and thiamine (Machlin, 1984).

Acetates, butyrates and ketogenic amino acids are utilized for energy by most tissues and organs, e.g. the heart (Bing, 1965). Important exceptions are the nervous system (Lindsay and Setchell, 1976), ovum, embryo and fetus which rely on glucose or the products of glycolysis.

The fatty acids pass through the membranes of most cells, except nervous cells. After cleavage and the formation of acetyl coenzyme A (CoA), they combine with oxaloacetate and enter and are oxidized in the TCA cycle and the terminal oxidative pathway. In the absence of sufficient oxaloacetate to incorporate all the available acetyl CoA into the TCA cycle, acetyl CoA is converted into acetoacetate, part of which is converted to β-hydroxybutyrate in a reversible reaction. Unlike the fatty acids, acetoacetate and β-hydroxybutyrate can enter nerve cells and can provide some but not all of the energy they require. Acetoacetate is converted into acetoacetyl CoA (taking the CoA from succinyl CoA) and thence to acetyl CoA in which form the acetyl groups are oxidized in the TCA cycle. No additional oxaloacetate is synthesized by the oxidation of fatty acids.

Whilst large quantities of substrates are required for and utilized by the cells for energy, only very small quantities of amino acids, vitamins and minerals are required for the glucose catalytic enzyme systems.

The concentration of glucose in the blood is determined mainly by the intake of glucogenic nutrients, the rate of gluconeogenesis and the rate of utilization for fetal development and milk synthesis.

The amount available for metabolism depends upon the concentration of glucose in the blood, the size of the glucose pool and the rate of entry into cells, with the latter being dependent upon insulin (except for nervous tissue).

Protein and Non-protein Nitrogen

Dietary non-protein nitrogen (NPN) and rumen-degradable protein (RDP) nitrogen are utilized by the ruminal bacteria for fermenting the complex carbohydrates in the cell-walls. The bacterial and protozoal protein synthesized in the rumen and reticulum, together with that part of the dietary protein which is not degraded, passes into the intestines. That which is hydrolysed and absorbed supplies amino acids for growth and metabolism or is deaminated. Some becomes available for gluconeogenesis. The quantity of amino acid N which passes into the duodenum is closely related to the dietary intake of DOM (Tamminga and von Hellemond, 1977).

Feeding diets of grass which have been heavily fertilized with N (and consequently have high NPN, protein and nitrate contents) increases NPN,

NH$_3$ and NO$_3$ concentrations in the ruminal liquor, lowers total volatile fatty acid (VFA) concentrations and increases the proportions of acetate (Bryant and Ulyatt, 1965).

The best rates of fermentation and microbial protein synthesis can be achieved only when the microorganisms are presented with sufficient fermentable carbohydrate and degradable protein or NPN. If the RDP and NPN are inadequate, the rate of microbial synthesis falls, decreasing the fermentation rate, the rate of passage of ingesta, appetite and food intake. If the rate of degradation of RDP to NH$_3$ exceeds the requirement for microbial synthesis there is waste and possibly toxic effect. In some cases the dietary fault is not an excess of NPN or RDP but a deficiency of readily available carbohydrate (mono- and oligosaccharides, fructosans and starch).

Paquay *et al.* (1973a,b) estimated that the optimal dietary digestible N : ME ratio ranges from 2.2 g N MJ^{-1} during the first three months of lactation to 1.7 g N MJ^{-1} ME during the sixth and seventh months and 1.3 g N MJ^{-1} during the tenth month. When expressed as a g N intake per MJ ME intake, the optimal ratio ranges from 1.6 after calving to 1.1 at the end of lactation. In the new UK Metabolisable Protein System (AFRC, 1992) the ratio of effective rumen-degradable protein (ERDP) to fermentable metabolizable energy (FME) is calculated, a typical value being 11 g ERDP MJ^{-1} FME.

The protein which is not degraded in the rumen (undegradable protein, UDP) passes through and is available for hydrolysis in the small intestine. The proportion of protein which is UDP varies with the species, e.g. fishmeal has a high proportion of UDP, pasture a low proportion. Feedstuffs containing a high percentage of UDP can provide more amino acids than other feedstuffs.

Minerals and Vitamins

Most of the minerals required by cattle for growth and metabolism are also required by the ruminal biota for the fermentation of the plant cell-walls and the release of the cell contents. They therefore play an essential role in digestion and thus influence appetite and feed intake.

The metabolism of six minerals: cobalt, copper, iodine, manganese, phosphorus and selenium, deficiencies of which have been reported to have caused infertility in cattle, are more conveniently discussed in Chapter 4.

The water soluble vitamins are normally synthesized by the ruminal biota and are unlikely to be deficient except where the bacterial metabolism has been inhibited, e.g. by the use of antibiotics.

β-Carotene and α-tocopherol are also considered along with the six minerals in Chapter 4.

Nutrient Requirements

Most countries with developed cattle industries have determined the nutrient requirements of cattle applicable to their systems of management. Systems in use in the UK (ARC, 1980, 1984; MAFF, 1984), the Netherlands, France, Germany, Scandinavia and the USA (NRC, 1978, 1984) have been compared by McDonald *et al.* (1991). There is no evidence to suggest that any one system is superior in determining the nutrient requirements for reproduction in the cow.

Special care is needed to ensure that the cow's requirements are met during the critical periods of growth to puberty, the last trimester of pregnancy and the early post-parturient period up to the time of implantation.

The requirements of lactating cows for maintenance and production are partly determined by the level of nutrition in late pregnancy (Hutton and Parker, 1973). If these are not met in early lactation, the cows draw upon their body reserves and consequently lose weight and condition (Wallace, 1955; Swanson and Hinton, 1962; Treacher *et al.*, 1986). Depending upon the condition at calving, concentrate-fed cattle may experience negative energy balances for up to three months after calving (Ferguson and Chalupa, 1989); pasture-fed cows even longer. Hutton (1963) observed that the intake of pasture did not reach its maximum until five months after calving.

Allowances need to be made for the extra energy which is required for grazing and exercise. These vary according to the pasture density, length and chemical composition (Wallace, 1956; Corbett, 1980). Makeham and Malcolm (1983) suggested plus 15% for feedlot cattle, plus 35% for cattle grazing medium to good pasture and plus 60% for cattle grazing poor pasture, with additional increments for disease, stress and extremes of temperature.

3

Occurrence, Prevalence and Incidence of Nutritional and Metabolic Infertility

Most chronic nutritional deficiences cause first the slowing of the growth rate, then in the adult the loss of body condition (muscle and subcutaneous fat) and ultimately emaciation, weakness and death. Reproduction fails at some stage in the course of the development of severe malnutrition. Production is also affected by malnutrition, which if sufficiently prolonged or severe, will affect the economy of the cattle enterprise. In these circumstances, the manager will be concerned more with growth rates, production and even survival than with reproduction. Infertility caused by malnutrition or 'nutritional infertility', therefore tends to be important mainly in cattle in unfavourable environments, and in cattle during and after unexpected adverse climatic conditions. The affected animals are mainly beef cattle, and, to a lesser extent, dairy herd replacement heifers and dry cows which, not being immediately productive, tend to be neglected in times of scarcity.

Cattle affected with metabolic diseases show few, if any, of the signs of malnutrition. When the metabolic failure occurs during the breeding season and involves the reproductive system, infertility can be manifest by no more than cows returning to oestrus after unsuccessful mating within a few days of the stress.

In the example shown in Fig. 3.1 the cows were changed from grazing autumn-saved to autumn-sown pasture just before the temporary reduction in first-service non-return rates during the second week of the spring mating season, then back again to autumn-saved pasture. No loss of condition was observed.

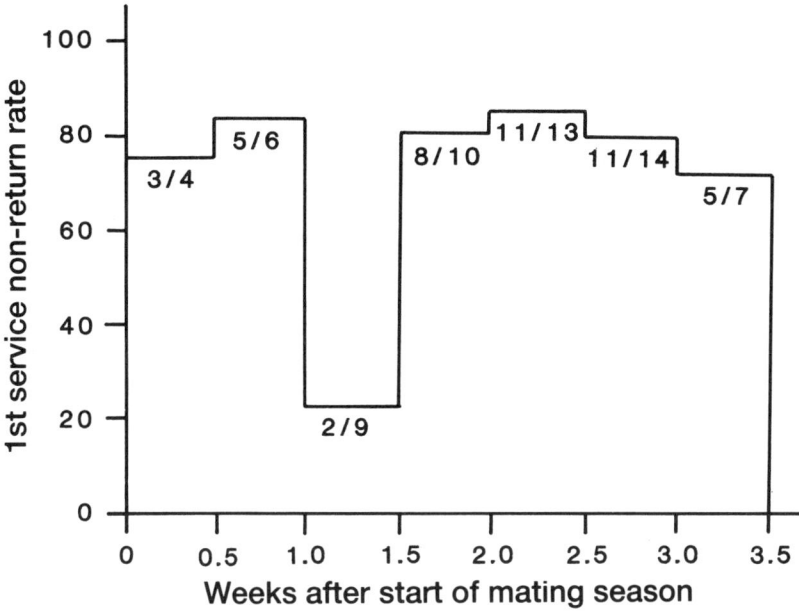

Fig. 3.1. Effect of a sudden change in pasture quality on first-service non-return rates (McClure, 1967c).

Occurrence

Nutritional and metabolic infertility can occur in cattle under some clearly defined circumstances:

1. Where the amount of food available is restricted during the critical stages of growth of heifers from weaning to puberty, during late pregnancy and during the early post-parturient period (lactation) up to and including the breeding season (see Tables 4.1, 4.2, 4.3 and 4.4).

Economic pressures and managerial errors cause most of these problems in hand-fed cows and heifers. Underfeeding in grazing cattle cannot always be controlled or even detected by herd managers. Often it results from unexpected climatic conditions such as drought, cold (Playne, 1969; Axelsen et al., 1972), prolonged wet and flood. Errors of management include judgements leading to overstocking, overestimation of length and density of pasture, insufficient grazing time, competition from sheep grazing the same pastures as the cattle and excessive distances between watering points.

Cows in commercial beef-breeding herds and even dairy herds grazing pastures in the less favourable climatic and geological areas, are sometimes underfed in late pregnancy. In such herds the effect on reproduction is greatest amongst those which are still growing, and is manifest in the heifers during their first lactation. The fertility of these heifers during their second breeding season tends to be lower than that of the adults which were not growing actively during pregnancy or the maiden heifers which were not lactating and hence were not subjected to the same degree of nutritional or metabolic stress. Young (1965, 1968) observed seasonal pregnancy rates of 68% in $L=01$ cows compared with 82% in $L=03$ cows and 87% in maiden heifers ($L=00$) in the East Coast district of New Zealand, and 80%, 93% and 84% respectively in New South Wales.

2. Mature grass/crop – where the cattle graze mature grass-dominant pastures or crops especially after the seed has fallen. The pastures most frequently involved are those that grow in the wet season in the high-rainfall zones of subtropical countries. Little rain falls in the winter and the grasses that grow during the summer wet season seed, dry off and remain as standing straw until they are grazed or knocked down and destroyed by the storm showers heralding the next wet season.

This dry native grass is poorly digestible and is deficient in protein, carotene and minerals including phosphorus (Franklin, 1959; Robinson and Sageman, 1967; Armstrong et al., 1968; Gartner et al., 1980, 1982).

Cattle grazing these pastures lose weight during the latter part of the dry season and the beginning of the next wet season. Both bodyweight and fertility recover when the summer rains come and feed quality improves (Barr, 1971). Lactating heifers perform worse than lactating cows which in their turn perform worse than non-lactating cows or heifers.

Cattle fed in a system which combines both hand feeding and pasture grazing may also suffer from a gross nutrient deficiency when the pasture is overmature and its low digestibility restricts the intake of digestible nutrients. This type of infertility occurs in dairy herds in subtropical, high-rainfall districts in the late summer when grass-dominant pastures which have run to seed form the major part of the pasture grazed by the milking herds. A summary of the analysis of the breeding records illustrating a herd problem of this type in New South Wales is shown in Table 3.1.

3. When lactating cows are fed lush young pasture or young forage crops. These feeds provide sufficient crude protein but insufficient energy and possibly rumen-undegradable protein (RUDP) to meet the requirements for maintenance plus milk production. Like the infertility that occurs in lactating cows fed on dry overmature grasses, the infertility in cows fed lush young grass or crops is accompanied by loss of weight. Losses of the order of 1% bodyweight per week have been reported associated with first-service non-return rates of 33% in lactating dairy cows fed *ad libitum* (McClure, 1970). Breeding records illustrating a problem of this type are shown in Table 1.4.

This type of feeding does not appear to be associated with infertility

Table 3.1. Reproductive performance of a dairy herd in New South Wales fed on overmature *Paspalum dilatatum*-dominant pasture.

	Dec	Jan	Feb	Mar	Apr[a]	May[a]	Jun[a]
No. of first services	2	11	4	2	17	10	7
Mean no. of days from calving	62	82	80	74	76	139	77
No. of subsequent services	2	1	6	5	6	9	8
No. pregnant to first service	2	4	0	1	6	4	7
Pregnancy rate (%)	100.0	36.4	0.0	50.0	35.3	40.0	100.0
No. pregnant to subsequent services	1	0	3	2	5	9	6
Pregnancy rate (%)	50.0	0.0	50.0	40.0	83.3	100.0	75.0
Pregnancy rate (total) (%)	75.0	33.3	30.0	42.9	47.8	68.4	86.7
Services per pregnancy	1.3	3.0	3.3	2.3	2.1	1.5	1.2
Mean calving-to-conception interval (days)	41	77	97	139	113	156	80

[a] Winter species would have outgrown the paspalum in April, May and June.

when cows are also fed sufficient grain-based concentrate rations (McClure, 1970).

4. When the pastures or crops have been grown on soils which contain insufficient available minerals considered to be essential for reproduction. The minerals involved have been reported to include cobalt (Underwood, 1981), copper (Underwood, 1981), manganese (Wilson, 1966), phosphorus (Theiler and Green, 1932) and selenium.

5. When a large part of the diet consists of one species of plant (or product manufactured from that plant) and that plant contains toxic substances or is deficient in essential nutrients. Examples include:

(a) Lucerne (alfalfa) (*Medicago sativa*) (Lotan and Adler, 1960; Adler and Trainin, 1961), subterranean clover (*Trifolium subterraneum*) and red clover (*Trifolium pratenese*) (Thain, 1967) containing phyto-oestrogens.

(b) Kale (*Brassica oleracea*) which sometimes contains goitrogens and tends to be deficient in copper, phosphorus and manganese (Munro, 1957; Jones, 1959; Alderman, 1963).

(c) Maize (*Zea mays*) silage, which contains little carotene (Lotthammer et al., 1978) and often insufficient methionine and other essential amino acids and little selenium.

(d) Sugarbeet (*Beta vulgaris*) tops and pulp which often contain insufficient phosphorus (Munro, 1957; Alderman, 1963) and manganese (Jones, 1959).

6. When cows are fed on manufactured feeds which are not formulated or compounded according to acceptable standards. Feeds mixed on farms are most at fault. The range of nutritional errors possible is wide and those affecting reproduction possibly include feeding insufficient forage or a low roughage to concentrate ratio (Buchanan-Smith et al., 1964; Franzos, 1968; Francos et al., 1977; Davidson et al., 1978; Bogin et al.,1982). Probably the most common error is to overestimate the quality or the intake of the roughage component of the ration.

7. When heifers are growing at subnormal rates and cows calve in poor condition or are losing weight at the time of mating. Liveweight and condition are considered in detail amongst the causes of reproductive failure (Chapter 4).

8. When cattle are fed on crops or pastures which are grown on soil containing high concentrations of molybdenum or are contaminated by industrial pollution (Pawlina et al., 1988; Simonik and Pavelka, 1989) or when cattle drink water containing high concentrations of fluorine (van Rensburg and de Vos, 1962).

9. When cows calve in fat condition.

10. When cows are fed excess rumen-degradable protein.

Prevalence and Incidence

There are few modern data indicating the prevalence of nutritional and metabolic infertility in either the beef or dairy herds of the various countries or even districts throughout the world. One reason for this is the ephemeral nature of the disease making it necessary to carry out prospective surveys of random samples of the population. Recent correlation studies suggest that nutritional and metabolic factors are responsible for a substantial proportion of the infertility in both beef and dairy herds; for example, Butler and Smith (1989) attributed much of the fall in pregnancy rates of New York State dairy herds from 66% in 1951 to 50% by 1973 and thereafter to energy deficiency induced by increasing milk yield.

4

Aetiology, Pathogenesis and Clinical Signs

Aetiology and Pathogenesis

At one time or other, deficiencies of almost all of the essential nutrients have been considered to cause infertility in the cow (Meites, 1953). While deficiencies of most have been shown to be capable of causing infertility in laboratory rodents, few have been proven to do so in cattle. With time, opinions concerning the nutrients responsible for reproductive failure in the field have changed. This may reflect changing nutritional and breeding practices, improved genetic merit of the cattle, or the better experimental techniques now used in studying the causes and pathogeneses.

The pathogenesis of nutritional and metabolic infertility is still far from clear. Reproductive failure can occur at three stages: (i) pituitary synthesis/release of luteinizing hormone (LH); (ii) ovarian function; or (iii) ovulation, fertilization and development of the ovum, embryo and fetus.

The evidence from cattle, sheep and laboratory rodent experiments indicates that energy deficiency can inhibit reproduction at all three stages. The pathogenesis of the failure caused by protein deficiency and 'excess' non-protein nitrogen/rumen-degradable protein is unclear. Similarly the pathogenesis of the infertility caused by mineral and vitamin deficiencies or excesses is not well understood.

Some cobalt, copper, iodine, manganese and phosphorus are involved in energy metabolic pathways; copper, selenium, α-tocopherol and β-carotene are anti-oxidants and may have a role in protecting the ovum, embryo and fetus and possibly other tissues from dietary polyunsaturated fatty acid peroxidation.

Reproductive failure (mediated by nutritional and metabolic means) may occur secondary to diseases of other systems. Of importance are diseases of the gastrointestinal tract, chronic systemic bacterial, parasitic and toxicological diseases and diseases causing liver dysfunction. Such diseases are identified by their clinical signs referable to the other systems involved.

Two causes of fatty degeneration of the liver, which have been described as the fatty liver syndrome and ketosis, have a nutritional basis and are referred to on pp. 45, 54.

Bodyweight and condition

Reproductive performance of cows is often but not invariably associated with bodyweight, bodyweight change and condition.

The growth rate of heifers seems to determine the age at which they reach puberty for this occurs at a bodyweight appropriate for the breed, and possibly strain (Table 4.1).

Subsequently, severe bodyweight loss is usually accompanied by anoestrus (Bond et al., 1958; Richards et al., 1989a). No precise critical weight-loss data are available, or probably even possible.

Recurrence of oestrous cycles after calving is associated with bodyweight change in late pregnancy and body condition at the time of calving. Cows in moderate to good condition (condition score of ⩾2.5 in the range of 1–5) return to oestrus in the minimal time; those with lower scores, or which have lost maternal bodyweight in late pregnancy, take progessively longer (Whitman, 1977; Wright et al., 1992a,b; Tables 4.2, 4.3; Fig. 4.1).

Loss of weight after parturition seems to delay the recurrence of oestrus, but to a much lesser degree than weight and condition changes before, and condition at, calving (Table 4.4). Wettemann (1982) recorded a delay of 19 days for every 10% loss of bodyweight.

Table 4.1. Effect of plane of nutrition and bodyweight on age at puberty in Holstein heifers.

Plane of nutrition (as % of recommended standard)	Bodyweight at puberty (kg)	Age at puberty (days)	Reference[a]
Low	61	241	504	(1)
	60–70	239	474	(2)
	62	289 ± 40	616	(3)
Medium	93	271	344	(1)
	100	265 ± 29	343	(3)
High	129	271	262	(1)
	110	257	372	(2)
	146	278 ± 32	280	(3)

[a](1) Sorensen et al. (1959); (2) Crichton et al. (1959); (3) Reid et al. (1964).

Table 4.2. Effect of plane of nutrition during late pregnancy on post-partum reproduction.[a]

Plane of nutrition pre-partum	Liveweight change (kg) Pre-partum	Liveweight change (kg) Post-partum	Mean calving-to-first-oestrus interval (days)	% cycling	First-service pregnancy rate	% not pregnant[b]	Reference
Low	−53.6	−16.4 ⎱ by	65	85 ⎱ by	65	0	Wiltbank et al. (1962)
High	+30.5	−61.4 ⎰ D90	48	95 ⎰ D90	67	5	
Low	−70.5				(Weaning rate 75.4%)		Hight (1966)
High	−17.3				(Weaning rate 93.4%)		
Low	−36.4					3	Hight (1968)
High	+26.4					9	
Low			70			25	Turman et al. (1964)
High			56			7	
Low	+8	+118 ⎱ by		88 ⎱ by	73	10	Dunn et al. (1969)
High	+68	+98 ⎰ D120		93 ⎰ D80	63	17	
Low	−0.5 ⎱	+0.83 ⎱	108		85		Kroker and Cummins (1979)
Medium	0 ⎰ daily	+0.62 ⎰ daily	92		45		
High	+0.75	+0.41	48		59		

[a] Cows were fed on a high plane post-partum.
[b] At end of mating season.

Table 4.3. Effect of plane of nutrition during late pregnancy and early lactation on reproduction.

Plane of nutrition		Liveweight change (kg)		Mean calving-to-first-oestrus interval (days)	% cycling	First-service pregnancy rate	% not pregnant[a]	Reference
Pre-partum	Post-partum	Pre-partum	Post-partum					
Low	Low	−60.0	−87.7 } by D90	52	22 } by D90	33	33	Wiltbank et al. (1962)
High	High	+30.5	−61.4	48	95	67	5	
Low	Low	−36.4	+0.22 } daily			45	45	Hight (1968)
High	High	+26.4	+0.38			9	9	
Low	Low	+8	+60 } by D120		73 } by D80	53	27	Dunn et al. (1969)
High	High	+68	+98		93	63	17	
Low	Low then medium					46	31	Wiltbank et al. (1964)
	Medium					54	29	
	Medium					31	22	
	High then medium					83	8	
	Low then high					87	0	

[a] At end of the mating season.

Table 4.4. Effect of plane of nutrition during early lactation on reproduction.[a]

Plane of nutrition post-partum	Liveweight change (kg)		Mean calving-to-first-oestrus interval (days)	% cycling		First-service pregnancy rate	% not pregnant[b]	Reference
	Pre-partum	Post-partum						
Low	+40.5	−130.5 } by	43	86 } by	42	16	Wiltbank et al. (1962)
High	+30.5	−61.4 } D90	48	95 } D90	67	5	
Low	+68.0	−28.0 } by		81 } by	56	36	Dunn et al. (1969)
High	+68.0	+98.0 } D120		93 } D80	63	17	
Low		−21%			60	18	Sommerville et al. (1979)
Medium		−16%			80	8	
High		−8%			84	6	
					s:c ratio	c-c interval (days)[c]		
Low		−0.06 } daily			1.94	100	Grainger and Wilhelms (1979)
High		+0.13 } daily			1.57	82	
Low			80			22	Smeaton et al. (1983)
High			78			7	

[a] Cows were fed on a high plane pre-partum.
[b] At the end of the mating season.
[c] s:c = service/conception; c-c = calving-to-conception.

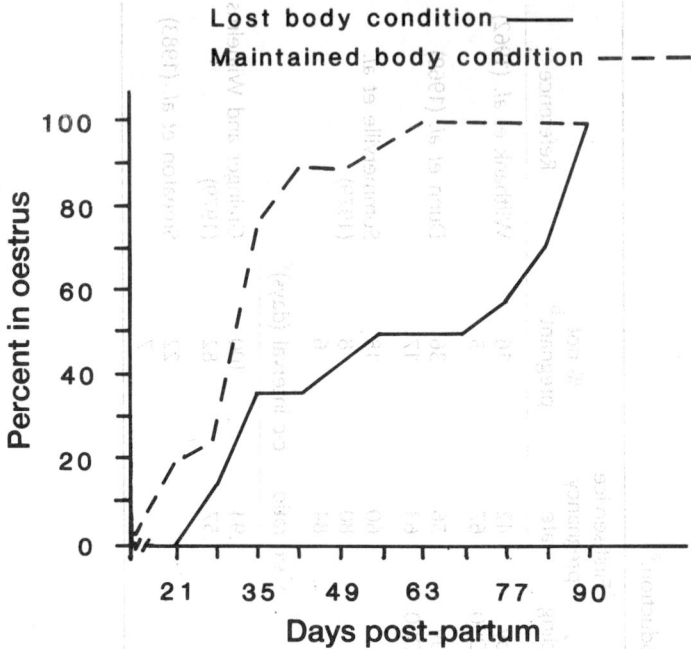

Fig. 4.1. Effect of body condition on the interval from calving to first oestrus (Rutter and Randel, 1984). (Reproduced by permission of the Editor, *Journal of Animal Science*.)

Low pregnancy rates have been reported in dairy cows that had lost excessive amounts of weight in the early part of lactation (>10%; Heinonen *et al.*, 1988) or were still losing weight at the time of mating >1% per week (McClure, 1961) and beef cows which were regaining weight only slowly (<14 kg; Greathead, 1983) at the time of mating. Dairy cows in poor condition and beef cows with condition scores of <2.5 (scale 1–5) at the time of mating have also been reported to be infertile (Leaver, 1977; Kilkenny, 1978, 1982; Rae *et al.*, 1993; Fig. 4.2).

Thus fertility is commonly correlated with bodyweight, bodyweight change and condition (i.e. muscle, intramuscular fat and subcutaneous fat). It remains to be seen whether this relationship is causal or casual and it is unlikely that this will be known until the pathogenesis of nutritional and metabolic infertility is fully determined. Both are manifestations of the nutritional state and are influenced by partitioning of nutrients and differential demands for muscle and fat anabolism, pregnancy and lactation.

In the short term at least, liveweight change and fertility can be disassociated by using the appropriate techniques (McClure, 1970; Downie and Gelman, 1976).

This separation of fertility and bodyweight change could explain the

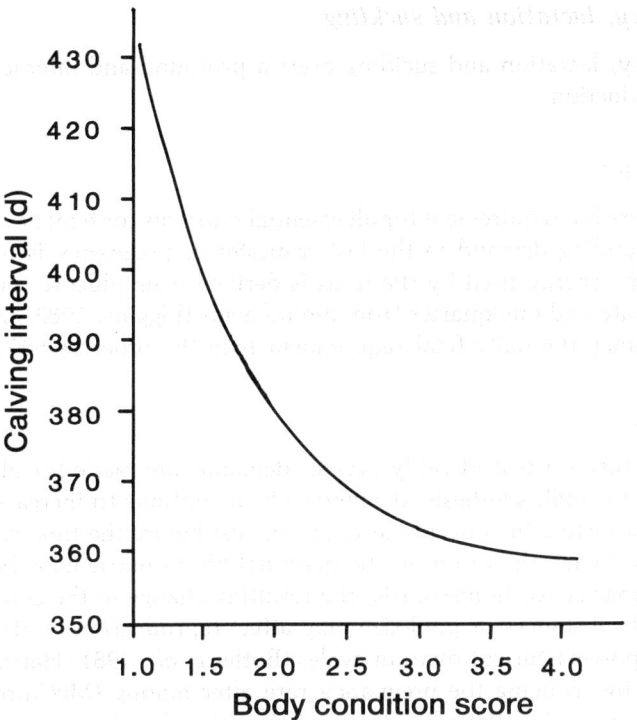

Fig. 4.2. Relationship between body condition score in beef cows at mating and subsequent calving interval (after Kilkenny, 1978). (Reproduced by permission of International Publishing Enterprises, Rome.)

discrepancies noted by Broster (1973) when reviewing liveweight change and fertility in the lactating cow and makes it unlikely that critical values will be determined which can be used to predict reproductive performance from weight change and condition-score data in individual cows. The measures are simply too crude to be of value in indicating metabolic state. Herd or group fertility is better correlated and therefore more predictable from liveweight and condition-score data.

Cows which calve in fat condition eat less during early lactation, mobilize more body tissue and lose more weight after calving than those calving in moderate condition. There is some evidence suggesting that these cows are less fertile and more susceptible to other diseases (Treacher *et al.*, 1986; Lotan *et al.*, 1988) including the 'fat cow or fatty liver syndrome' (Reid, 1980, 1983) and the 'parturition syndrome' (Sommer, 1975).

Genetic merit plays some part in the liveweight and condition changes which can occur during lactation, with superior cows producing more milk and losing more weight and condition (Holmes and Macmillan, 1982).

Pregnancy, lactation and suckling

Pregnancy, lactation and suckling exert a profound and interacting effect on reproduction.

Pregnancy

While there is a requirement for all essential nutrients for fetal development, the outstanding demand in the last trimester of pregnancy is for energy. Half of the energy used by the fetus is derived from glucose, one quarter from lactate and one quarter from amino acids (Liggins, 1982). At the end of pregnancy the daily fetal requirement is in the order of 500 g glucose.

Lactation

After parturition considerably greater demands are made for all essential nutrients for milk synthesis, demands which continue to increase until the peak milk yield which occurs at about or just before the time cows are to be mated. When the drain on the maternal blood nutrients is beyond the body's capacity for homeostasis, the resulting change in the concentration of crucial substances or pool-size may affect reproduction by delaying the onset of post-parturient ovarian cycles (Butler et al., 1981; Harrison et al., 1989) or by reducing the pregnancy rate after mating (McClure, 1970).

Ducker et al. (1985a,b) identified the critical factor causing the low pregnancy rate as being the degree of net-energy output at the time of mating and found that the fertility was inversely correlated with milk yield, change in milk yield and the yield of protein and lactose. The administration of bovine somatotrophin, while increasing milk yield, decreases both condition score and fertility (McClary et al., 1990).

Suckling

On its own, suckling, though delaying the release of gonadotrophin-releasing hormone (GnRH) from the hypothalamus after calving by stimulating β-endorphin release (Oxenreider, 1968; Oxenreider and Wagner, 1971; Lamming et al., 1981; Smith et al., 1981; Gordon et al., 1987; Nett, 1987; Connor et al., 1990), does not do so sufficiently to prevent inter-calving intervals exceeding 365 days. But it does provide an added component to the nutrition–lactation interaction. In dairy cows where the high milk yield causes a large drain on nutrients and metabolites, it is the interaction between nutrition and lactation which is important. In beef cows it is the interaction between nutrition, lactation and suckling which becomes critical (Oxenreider and Wagner, 1971).

These interactions delay the recommencement of oestrous cycles after parturition and lower pregnancy rates.

Minerals

Diets which contain low concentrations of cobalt, copper, iodine, manganese, phosphorus and selenium or high concentrations of molybdenum, have been reported and are commonly believed to cause infertility in cows as they have been shown to do in laboratory rodents. Some of the earlier claims were based on comparisons of fertility before and after the diets had been supplemented with minerals, a technique now recognized as being unreliable. Others used supplements which were inadequately defined in respect to their micronutrient composition. Many of the reports also referred to work done before it was realized that energy deficiency/ imbalance is a major cause of nutritional infertility. Finally, most of the work reported was done before it was realized that metabolic infertility can result from acute deficiencies or imbalances which may act for periods which are too short to cause visible signs of malnutrition and which can alter during the course of experiments. The role of mineral deficiencies in the aetiology and pathogenesis of bovine infertility needs reinvestigation, keeping energy intakes high enough to meet the animals' requirements for maintenance, growth, pregnancy and milk synthesis and protein levels adequate for maximal digestibility and appetite.

The pathogenesis of the resulting reproductive failure is likely to involve:

1. Depression of the activity of the ruminal microbiota and consequently reduced rate of fermentation, appetite and intake of digestible organic matter (DOM). The responses could therefore occur within hours of the ingestion of the deficient ration, though this is more likely to be delayed because of the contribution saliva makes to the ruminal liquor.
2. Reduction in the rate of enzymatic action affecting energy and protein metabolism and synthesis of hormones. As the quantities of the minerals required for this are small and recycling occurs, any infertility is likely to result from chronic rather than acute deficiencies.
3. Interference with the synthesis or integrity of cells in the reproductive system, in particular those which are rapidly dividing, such as in the endometrium, ovum, embryo and fetus.

Imbalance between the concentrations of competing minerals in the feedstuffs (such as iron, molybdenum and selenium) may interfere with the absorption or retention of critical minerals like copper (Phillippo *et al.*, 1982, 1985; Anke *et al.*, 1991; Davies *et al.*, 1992).

Cobalt

Severe chronic deficiencies of cobalt were common in pasture-fed cattle in many parts of the world including Australia, Florida, Kenya, New Zealand

and Scotland up to the late 1930s when the cause of the ill-thrift and emaciation in these areas was discovered and control methods developed (Filmer, 1933; Filmer and Underwood, 1934, 1937). Under these conditions of severe deficiency the dominant clinical signs of inappetance, failure to grow and fatten, ill-thrift and loss of condition leading to emaciation and death were of greater and more urgent importance than reproduction. Infertility has received scant mention in the literature though almost certainly it would have occurred. The control measures have been highly effective and clinical cobalt deficiency is now almost unknown. What does concern us is the possible effects of mild degrees of cobalt deficiency.

Cobalt deficiency results from feeding diets containing <0.07 mg Co kg^{-1} dry matter (DM), preventing the formation of cyanocobalamin (vitamin B12) (Marston et al., 1961) and thus coenzyme B12 required for the conversion of propionate to succinate. This reduces gluconeogenesis and the activity of the tricarboxylic acid (TCA) cycle, and results in hypoglycaemia (MacPherson et al., 1973).

Cobalt deficiency may also reduce the amount of adenosyl cobamide coenzyme which is necessary for the incorporation of deaminated glucogenic amino acids into the TCA cycle.

Copper

Copper is widely distributed in the body in the enzymes cytochrome oxidase, copper-superoxide dismutase, dopamine-β-hydroxylase, monoamine oxidase, ceruloplasmin and lysyloxidase (Howell and Davison, 1959; Nagatsu, 1973; Mason, 1979).

Clinical copper deficiency occurs when cattle are fed diets containing <3 mg Cu kg^{-1} DM when the molybdenum (Mo) content is <3 mg kg^{-1}, 3–10 mg Cu kg^{-1} when the Mo content is 3–10 mg kg^{-1}, and >10 mg Cu kg^{-1} when the Mo content is >10 mg kg^{-1}.

Infertility has been observed associated with clinical signs of copper deficiency in many parts of the world (Underwood, 1981). At these rates of copper and molybdenum intake, the efficiency of food utilization is reduced (Mills et al., 1976), probably because of cellular damage to the small intestine affecting absorption (Fell et al., 1975). Field evidence suggests that copper deficiency can affect reproduction in the absence of other clinical signs except perhaps for achromotrichia (Munro, 1957, 1975; Hunter 1976). The significance of these early reports has been questioned by the work of Phillippo et al. (1982, 1985) who found excessive concentrations of molybdenum in the feed of infertile cows with low serum copper concentrations which failed to respond to copper therapy.

Copper deficiency may cause infertility because of its effect on the integrity of the small intestine (Fell et al., 1975, Mills et al., 1976) which would lead to a general nutrient deficiency, and its roles in energy metabolism and as an anti-oxidant. Data from the laboratory rat show that copper-deficient diets cause primary fetal death between day 12 (D12) and D13 (Howell and Hall, 1969; Beguin et al., 1985).

Iodine

Iodine deficiency occurs as the result of feeding pasture or crops grown on iodine-deficient soils containing <2 mg I kg^{-1} DM or excessive quantities of goitrogens such as thiocyanates in the feedstuffs. Cattle fed on such crops have been reported to be infertile with affected cattle exhibiting irregular or suppressed oestrus, and the calves being aborted, born weak or dead or affected with goitre and fetal membranes retained (Allcroft et al., 1954; McDonald et al., 1962; Alderman, 1963; Danilin, 1966; Underwood, 1977; Ryot et al., 1990).

There is no evidence to suggest that iodine-deficiency infertility, either absolute or induced by goitrogens, is anything but rare.

Apart from its role in preventing thyroid hyperplasia, iodine forms part of the thyroxine molecule which stimulates mitochondrial ATP production. Deficiency reduces the oxygen uptake by cells indicating a reduction in the rate of energy metabolism (Peterson et al., 1952, Underwood, 1977). It is not known whether this occurs sufficiently severely as to cause reproductive failure in cattle as has been reported in rabbits (Peterson et al., 1952).

Manganese

Manganese deficiency has been reported in cows fed on diets containing <22 mg Mn kg^{-1} DM and appears to cause anoestrus or suboestrus, delayed ovulation, low first-service pregnancy rates and the birth of deformed calves (Bentley and Phillips, 1951; Wilson, 1952, 1965, 1966; Hignett, 1956, 1959; Munro, 1957; Dyer and Rojas, 1965; Rojas et al., 1965; Krolak, 1968; Rasbech, 1968). The infertility occurs mainly in cattle grazing pasture growing on manganese-deficient or heavily limed soils or fed supplements containing excessive amounts of calcium. Being a cofactor for, or a component of, the enzymes phosphoglucomutase, pyruvate carboxylase and phosphoenolpyruvate carboxyl kinase, manganese has a key role in gluconeogenesis (Baly et al., 1985).

Phosphorus

The role of phosphorus in the pathogenesis of nutritional and metabolic infertility is unclear. Early reports associated dietary concentrations of <2.0 g P kg^{-1} DM with reproductive failure (Tuff, 1923; Theiler and Green, 1932; Eckles et al., 1935; O'Moore, 1950; Alderman, 1963, 1970). The feedstuffs incriminated were pasture and hay made from pastures grown on phosphorus-deficient soils, mature grass, and hay made from mature grass, and species of plants which naturally contain low concentrations of phosphorus, e.g. kale and sugarbeet (Munro, 1957; Alderman, 1963).

Forages which are deficient in phosphorus are often deficient in available carbohydrate, protein, other minerals and carotene (Eckles et al., 1935; Palmer et al., 1941; Franklin, 1959; Gartner et al., 1980) making the assessment of the significance of the published work difficult.

Phosphorus is an important constituent of bone, and of phosphatic esters, nucleotides, coenzymes and inorganic orthophosphate. These are involved in many metabolic processes in the body, particularly those in energy metabolism. The most obvious of these is its contribution to the $ATP \rightarrow ADP + \sim P$ reaction.

Any reduction in the availability of phosphorus for cellular metabolism is likely to restrict energy flow.

In attempting to determine the role of its deficiency on reproductive function we need to consider:

1. The depressant effect on digestibility, appetite and intake (Long et al., 1957; Little, 1968, 1970; Gartner et al., 1982).
2. The interaction between phosphorus intake and lactational drain (McTaggart, 1959; Hart and Mitchell, 1965; Hunter, 1976).
3. The potential for replenishment of phosphorus in the tissue fluids and saliva from bone.

The failure of Cohen (1975), Hecht et al. (1977) and others to find any relationship between phosphorus intake and fertility underlies the necessity for more properly conducted trials.

Other minerals

The potential anti-oxidant role of copper is considered on p. 59, together with selenium and α-tocopherol.

The ions Na^+, Ca^{2+} and Mg^{2+} play critical roles in regulation of cellular function and although they are involved in the pathogenesis of other metabolic diseases (Payne, 1977) they have as yet no confirmed role in metabolic infertility in the cow. However, they are potentially important; e.g. Block (1991) referred to the role of ATP in supplying energy for the Na^+-K^+ pumping mechanism and Rega and Garrahan (1986) for the Ca^{2+} pump.

Energy

The most common and most important nutritional cause of reproductive failure in the cow is the failure of the digestive and hepatic systems to provide sufficient energy for maintenance, exercise, growth, pregnancy and lactation. Reproduction can be affected at any stage up to the time of implantation; the stage and the lesions produced being dependent upon timing and degree of the energy deficit.

Factors involved in the aetiology and pathogenesis of infertility induced by energy deficiency in the cow are detailed below.

Plane of nutrition

Underfeeding during growth delays puberty; during late pregnancy delays the recurrence of oestrous cycles after parturition; and during the post-partum period reduces pregnancy rates.

The effects of underfeeding during growth to puberty are summarized in Table 4.1, during the late-pregnant period in Table 4.2, the late-pregnant–early-lactation period in Table 4.3 and the early-lactation period in Table 4.4. Blood-glucose concentration, pool-size and the rate of entry of glucose into cells is reduced (Kronfeld and Raggi, 1964) and the pulsatile release of LH by the pituitary is suppressed (Randel, 1990).

As economic factors dictate that cattle be fed the minimal quantity of food consistent with achieving the production goals, there is always some risk that the amount of feed made available will fall below the optimum to such an extent that fertility suffers. This general principle holds true whether the major feed is pasture (where the index of efficiency is yield per hectare of pasture land) or whether the main feedstuff is purchased (and the index is yield per cow).

Feed composition

The chemical composition of the feed influences the quantity of energy which is absorbed and also the nature of the energy substrates, whether they are glucogenic or ketogenic. This appears to have a causal role in the pathogenesis of nutritional and metabolic infertility. The critical components are: (i) the available carbohydrate content; (ii) the protein content and whether it is rumen-degradable or undegradable; and (iii) the mineral content.

1. **Available carbohydrate.** Manufactured grain-based feedstuffs should provide an adequate ruminal propionate : acetate ratio. Even the best-formulated feedstuff may not contain enough energy to supply the needs of cows producing high yields of milk or those treated with bovine somatotrophin (Hart *et al.*, 1978; McClary *et al.*, 1990) because of the constraints imposed by bulk and rumen-fill.

Pastures and crops contain low concentrations of available carbohydrate when very young and again when approaching maturity. The critical

DM percentage indicating these stages of maturity in forage oats were reported to be <16% and >20% with a minimal soluble carbohydrate plus starch concentration approaching 20% (Fig. 4.3). Lactating cows fed on either immature or overmature oats were found to have lower blood-glucose concentrations and to be less fertile than those fed oats within a DM range of 16–20% and ⩾20% soluble carbohydrate plus starch (McClure, 1970). Further, there is an interaction between milk secretion and feed composition on the blood-glucose concentration (Table 4.5; Fig. 4.4).

While restricting the quantity of a balanced ration to lactating beef cows does not seem to greatly affect the blood-glucose concentration (McClure, 1977a; Blum et al., 1979), unbalanced diets made up of different proportions of maize meal, cottonseed meal, urea and rice straw, deficient in readily available carbohydrate and/or protein are unable to maintain normal concentrations of glucose even when fed *ad libitum* (McClure, 1977a).

Some confirmation of this was provided by Shrick et al. (1990) who did not alter plasma cortisol, propionate, butyrate, cortisol, LH or progesterone concentrations by feeding a cracked corn, soyabean, Bermuda grass diet in quantities causing a daily loss of 1.37 kg bodyweight.

2. **Protein and non-protein nitrogen.** Insufficient dietary non-protein nitrogen (NPN) and rumen-degradable protein (RDP) reduce the digestibility of the feed. In turn, this causes a dietary energy deficiency and reduces the flow of microbial protein (Orskov, 1982), slows the growth rate of the heifer, delays or inhibits puberty and the commencement of oestrous cycles after calving, and reduces the pregnancy rate to first and subsequent services (Witt et al., 1958; Bedrak et al., 1964; Wiltbank et al., 1965;

Table 4.5. Effect of lactation and feed composition on dry matter (DM) intake (kg 100 kg^{-1} liveweight) and blood-glucose concentration (mmol l^{-1}) of cows fed forage oats (*Avena sativa*) of different stages of maturity. (McClure, unpublished.)

Stage of maturity	Lactating cows				Non-lactating cows			
	Glucose			Dry matter intake	Glucose			Dry matter intake
	n[a]	X̄	SD		n[a]	X̄	SD	
Immature <16% DM	18	1.33	0.18 (a)	2.1	18	2.75	0.41 (b)	1.7
Moderately mature 16–20% DM	18	2.04	0.31 (c)	3.0	18	2.91	0.17 (d)	1.7

[a] n = number of samples tested. There were three cows in each group.
(a) vs. (b) $P = <0.001$; (a) vs. (c) $P = <0.001$; (c) vs. (d) $P = <0.001$; (b) vs. (d) $P = >0.05$.

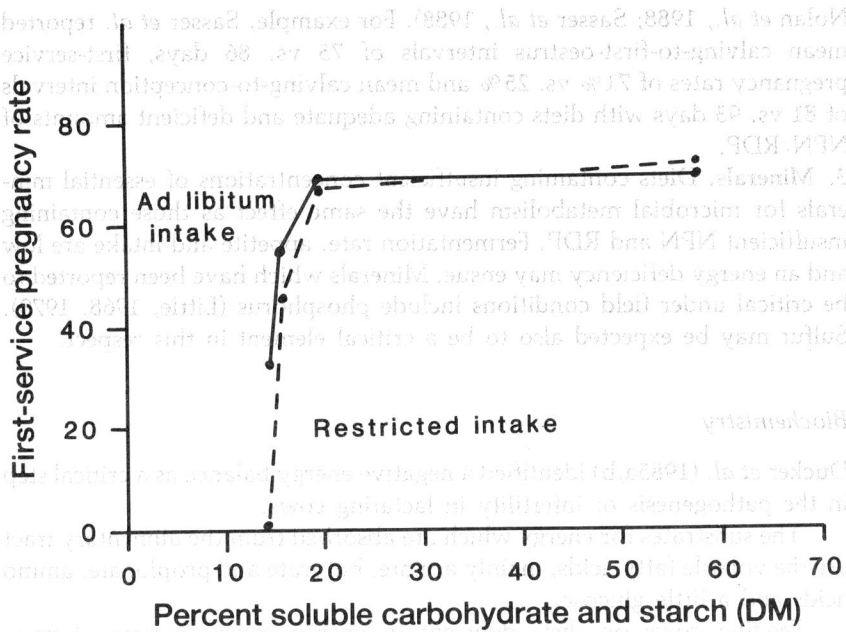

Fig. 4.3. Relationship between the first-service pregnancy rate of cows and the soluble carbohydrate and starch content of the feed (McClure, 1970).

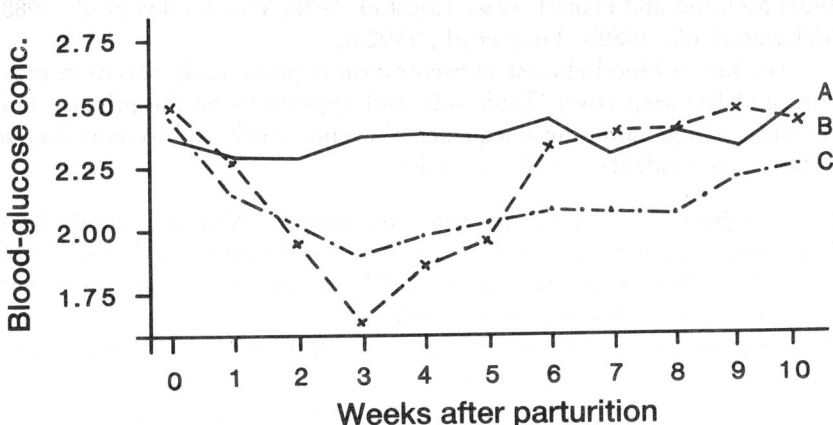

Fig. 4.4 Blood-glucose concentration in cows fed on different rations during early lactation. Ration A, forage oats of <16% DM plus hay and dairy meal. Ration B, forage oats of 16–20% DM. Ration C, forage oats of >20% DM plus dairy meal (McClure, 1977b). (Reproduced by permission of the CSIRO Editorial and Publishing Unit, Melbourne, Australia.)

Nolan *et al.*, 1988; Sasser *et al.*, 1988). For example, Sasser *et al.* reported mean calving-to-first-oestrus intervals of 75 vs. 86 days, first-service pregnancy rates of 71% vs. 25% and mean calving-to-conception intervals of 81 vs. 93 days with diets containing adequate and deficient amounts of NPN/RDP.

3. Minerals. Diets containing insufficient concentrations of essential minerals for microbial metabolism have the same effect as those containing insufficient NPN and RDP. Fermentation rate, appetite and intake are low and an energy deficiency may ensue. Minerals which have been reported to be critical under field conditions include phosphorus (Little, 1968, 1970). Sulfur may be expected also to be a critical element in this respect.

Biochemistry

Ducker *et al.* (1985a,b) identified a negative energy balance as a critical step in the pathogenesis of infertility in lactating cows.

The substrates for energy which are absorbed from the alimentary tract are the volatile fatty acids, mainly acetate, butyrate and propionate, amino acids and a little glucose.

Feeding cows on diets deficient in energy causes plasma glucose, insulin, growth hormone, LH, oestrogen and progesterone concentrations to fall and non-esterified fatty acids (NEFA) and β-hydroxybutyrate concentrations to rise (Donaldson *et al.*, 1970; Hill *et al.*, 1970; Kazmer *et al.*, 1985; McCann and Hansel, 1986; Ropstad, 1988; Villa-Godoy *et al.*, 1988; Richards *et al.*, 1989b; Lucy *et al.*, 1992a).

The fall in blood-glucose concentration is particularly severe in pregnant and lactating cows (Table 4.5) and appears to be the primary biochemical change initiating a sequence of events which lead to reproductive failure. The evidence for this includes:

1. Hypoglycaemic and ketotic cows are infertile (McClure, 1968, 1970; Oxenreider and Wagner, 1971; Hunter, 1976; McClure and Payne, 1978; Jones *et al.*, 1988) particularly when the blood-glucose concentration is falling (Downie and Gelman, 1976; Table 4.6).
2. Administration of insulin (McClure, 1968; Table 4.7) and the glucose metabolic inhibitor 2-deoxy-D-glucose (McClure *et al.*, 1978) cause reproductive failure. Acute hypoglycaemia induced by the administration of phlorizin altered the pattern of LH release in the post-partum beef cow (Rutter and Manns, 1987) and suppressed the normal increase in LH pulse magnitude during the pre-ovulatory period in cyclic cows (Rutter and Manns, 1988).
3. Restricting the intake of balanced feedstuffs within reasonable limits, even though causing appreciable losses in bodyweight, did not affect either the blood-glucose concentration or fertility (McClure, 1970, 1977a,b).

Table 4.6. Mean plasma glucose concentrations (mmol l^{-1}) ± SD, and liveweights (kg) of Hereford × Friesian cows, from two weeks before service to one week after service. (After Downie and Gelman, 1976.)

	−2 weeks	−1 week	Service	+1 week
Fertile cows ($n=14$)[a]				
Glucose	3.20 (±0.49)	3.43 (±0.46)	3.64 (±0.34)	3.52 (±0.36)
Weight	479.50	476.48	467.48	462.57
Infertile cows ($n=7$)[a]				
Glucose	3.34 (±0.30)	3.24 (±0.31)	2.81 (±0.51)	3.44 (±0.77)
Weight	468.34	457.31	451.41	449.45

[a] n = number of samples tested.

4. Feeding cows monensin increases the molar proportion of propionate at the expense of acetate; increases the blood-glucose concentration (Potter et al., 1976; Hixon et al., 1982); hastens the ability of the hypothalamus and pituitary to respond to oestrogens early in the post-partum period; reduces the age of heifers at puberty; and reduces the calving-to-conception interval (Baile et al., 1982; Randel et al., 1982a,b; Mason and Randel, 1983; Rutter et al., 1983; Sprott et al., 1988; Randel, 1990).

Acute energy deficiency and the administration of insulin and the glucose metabolic inhibitor 2-deoxy-D-glucose in rats and mice have been shown to affect reproduction at all stages of the cycle, producing all the lesions reported from infertile cattle from suppression of oestrus to embryonic death (McClure, 1966, 1967a,b). In a recent review, Short and Adams (1988) concluded that 'the only evidence for a mediator of the effects of feed level is blood-glucose concentration'.

Hypotheses as to the sequence of events consequent upon hypoglycaemia, the reduction in glucose pool-size or the rate of entry into cells include:

1. Inhibition of hypothalamic synthesis and/or release of GnRH due to:
    (a) Insufficient ATP for the GnRH secretory neurones and their synapses.
    (b) Stimulation of endogenous opioid secretory neurones.
Either would lead to the observed failure of the gonadotroph cells of the pituitary to secrete/release LH.
2. Inhibition of synthesis and/or release of follicle-stimulating hormone (FSH) and LH by the adenohypophysis due to:
    (a) Insufficient energy (glucose + insulin) for cellular metabolism.
    (b) Failure of GnRH stimulation because of insufficient numbers of GnRH receptors, or insufficient episodic or pulsatile GnRH.
3. Inhibition of the development of the follicle and ovum and the secretion of oestradiol, progesterone and inhibin due to:

Table 4.7. Effect of insulin on pregnancy rate (McClure, 1968). (Reproduced from the *British Veterinary Journal* by permission of the Editor.)

Treatment given on days	Pregnancy rate of insulin-treated and control cows		No. of cows pregnant/No. of cows mated	
	Cows mated 0–2 days after last injection	Cows mated 3–26 days after last injection	Total	
18, 19, 20 after last oestrus	1/9 ⎫ 2/11 18%	1/1 ⎫ 8/9 89%	2/10 ⎫ 10/20	50%
17, 18, 19, 20 after last oestrus	1/2 ⎭	7/8 ⎭	8/10 ⎭	38%
0, 1, 2, 3, after mating	–	–	6/16	38%
Nil, blood-sampled controls	–	–	15/20	75%

(a) Insufficient energy (glucose + insulin).
(b) Decreased sensitivity to gonadotrophins.
(c) Insufficient quantities of gonadotrophins.
4. Death of the ovum, embryo or fetus due to:
(a) Insufficient glucose (or glycolytic products) and probably insulin.
(b) Insufficient ovarian steroid hormones.

In rats hypoglycaemia inhibits the release of GnRH (Dyer and McClure, 1981) and the pulsatile release of LH (McClure and Saunders, 1985). Activation of an inhibitory opioid pathway may be involved in the pathogenesis in rats (Dyer et al., 1985) but though inducing suckling inhibition of LH release it does not appear to mediate the energy-deficiency inhibition of LH release or of ovarian response to LH in cows (Short and Adams, 1988; Canfield and Butler, 1991).

Randel (1990) concluded that in the cow the primary target for sensing and reacting to the nutritional status is the hypothalamus, and that hypothalamic release of GnRH is suppressed in cows fed diets low in energy. Pituitary gonadotrophic hormone secretion, differential sensitivity of the pituitary–hypothalamus to steroids and releasing hormones, and ovarian function seem to be affected (Short and Adams, 1988; Lucy et al., 1992a), but it is not clear whether this is the result of GnRH failure alone or whether the energy deficiency affects the pituitary, ovary, ovum, embryo and fetus directly.

The role of acetate and the free fatty acids (FFA) in the pathogenesis of nutritional and metabolic infertility have not been determined. The increase in plasma concentrations of FFA which occurs in lactating cows catabolizing fat should provide energy for pituitary and ovarian function, but not for the hypothalamus, ovum, embryo or fetus. For example, acetate can stimulate ovarian function in ewes in a manner and to a degree similar to glucose plus insulin (Teleni et al., 1989; Downing and Scaramuzzi, 1991), and calcium soaps of palm oil can increase LH and progesterone production in beef cows (Hightshoe et al., 1991).

Protein and non-protein nitrogen

Crude-protein concentrations in the diets of cows in the range of 13–20% can sustain normal reproduction (Carroll et al., 1988). Feeding cows from late pregnancy through to early lactation on diets containing insufficient RDP and NPN reduces appetite and food intake and has the same effect as underfeeding. Pituitary LH and FSH contents are reduced (Nolan et al., 1988), the recommencement of post-partum oestrous cycles is delayed or suppressed, ovulation may fail, and pregnancy rates are reduced and calving-to-conception intervals increased (Witt et al., 1958; Sasser et al., 1988).

The fertility of cows fed diets containing insufficient rumen-undegradable or 'by-pass' protein (UDP) may also be depressed (Wilson *et al.*, 1985; Ferguson *et al.*, 1986; Armstrong *et al.*, 1990). By reducing the proportion of RDP in an isocaloric diet containing 16% crude protein from 73 to 64%, Blanchard *et al.* (1990) increased the percentage of fertilized ova from 54.8 to 79.2%. The cause may be related to the fact that high-producing cows may also not absorb sufficient amino acids for both tissue metabolism and milk secretion (Tamminga and von Hellemond, 1977). In these circumstances the main limiting amino acids are likely to be methionine and/or lysine. This fact may be significant in the light of reports by Saiko and Krukovets (1976), Stoikov (1978), Kuzainov and Kozhabergenov (1988), Bonomi *et al.* (1989) and Spiekers and Pfeffer (1990) of therapeutic effects of methionine and lysine on fertility and a stimulatory effect on production (Griel *et al.*, 1968).

Alternatively, the role of the amino acids in the UDP might be to provide additional substrate for gluconeogenesis or the synthesis of fatty acids. In ewes the addition of lupins to the diet reduces the rate of atresia during the final stages of follicular development and increases the ovulation rate. The effects can be mimicked by the infusion of glucose, acetate (Teleni *et al.*, 1984, 1985, 1989; Kuswandi and Teleni, 1987) and the amino acids leucine, isoleucine and valine (Downing and Scaramuzzi, 1991).

High crude-protein concentrations in the diets of cows have been associated with infertility (Fernando and Carter, 1970; Girou and Brochart, 1970a; Jordan and Swanson, 1979a,b; Ferguson *et al.*, 1986; Ferguson and Chalupa, 1989). Any such effect may involve nitrate or ammonia toxicity (Bennet *et al.*, 1968) although the evidence is equivocal (Blauwiekel *et al.*, 1986) and the ammonia toxicity hypothesis, at least, appears to be untenable (Norton and Hogan, 1993). The effect, if any, may be due to either:

1. A depression in the total soluble sugar and water soluble carbohydrate content of the grass which is induced by heavy application of nitrogenous fertilizer to the pasture or lack of sunlight (Bryant and Ulyatt, 1965; Hogan and Weston, 1969; Fernando and Carter, 1970).
2. An effect of NO_3 on the carbohydrate fermentation by the ruminal flora encouraging the production of acetate at the expense of the glucogenic fatty acids (Bryant, 1965). The addition of urea to the feedstuff has been observed to cause blood-glucose concentration to fall (Leonard *et al.*, 1977)
3. The magnitude of the UDP component (Ferguson and Chalupa, 1989).

An alternative explanation of the infertility which has been associated with the feeding of high crude-protein diets containing high RDP : low UDP ratios is that such diets sometimes contain insufficient readily available carbohydrate for the rumen microbiota to fully utilize the free nitrogen produced from the RDP (see Fig. 4.3 above).

Anti-oxidants

Deficiencies of three or possibly four nutrients have in common equivocal data indicating that they are causes or potential causes of reproductive failure, that the failure appears to affect mainly the ovum, embryo or fetus, and that they are anti-oxidants, or are components of anti-oxidant systems. The nutrients are selenium, a fraction of glutathione peroxidase, α-tocopherol, β-carotene (Willson, 1983) and copper (copper-superoxide dismutase) (Diplock, 1983).

Of these, α-tocopherol and glutathione peroxidase are the most important and, by combining with the free radicals that initiate the oxygenation of unsaturated phospholipids and critical sulfhydryl groups, they help to inhibit lipid peroxidation and maintain the integrity of cell membranes (Rammell, 1983; Machlin, 1984; Putnam and Comben, 1987). α-Tocopherol is required particularly during fetal development. Rammell (1983) concluded that the ruminant requires tocopherol only to overcome the toxic effects of polyunsaturated fatty acids in the diet, or to prevent lipid peroxidation and membrane (both cellular and subcellular) damage in the absence of the far more efficient glutathione peroxidase.

Selenium and alpha-tocopherol

The best-known syndromes responding to selenium therapy are muscular dystrophy (Hidiroglou and Jenkins, 1968), poor growth rates (Gleed et al., 1983), abortion (Taylor et al., 1979; Ohba et al., 1992) and retention of fetal membranes (Trinder et al., 1969; Eger et al., 1985).

Limited evidence suggests the involvement of selenium and/or vitamin E in the pathogenesis of infertility (Segerson et al., 1977; Larson et al., 1980; McClure et al., 1986), but other reports (Southcott et al., 1972; Larson et al., 1980; Morrow et al., 1981; Harrison and Conrad, 1982; Ishak et al., 1983; Harrison et al., 1984; Stowe et al., 1988; Allan et al., 1993) indicate little or no effect on reproduction. Big between-herd differences in response have been reported (Jaskowski and Rogoziewicz, 1990; McGowan, 1991).

Larson et al. (1980) recorded an association between the blood-selenium concentration 14–21 days after parturition and the service : pregnancy ratio and the number of days between calving and conception. McClure et al. (1986) increased the first-service pregnancy rate from 30 to 58% in a controlled field trial using selenium and iron pellets administered per os in dairy herds with mean blood-glutathione-peroxidase (GSH-Px) activity <70 units g^{-1} haemoglobin. The field evidence reported by McClure et al. (1986) which showed a lengthening of the oestrous cycles after unsuccessful service suggests the possibility of embryonic mortality as occurs in sheep (Hartley, 1963). The conflicting evidence suggests the possibility of interaction with other undefined factors.

Selenium deficiency can occur where cattle are fed on pasture, crops or grain from crops grown on soils containing <0.5 mg Se kg^{-1} DM, and feedstuffs containing <0.05 mg Se kg^{-1} DM. α-Tocopherol deficiency can occur in cattle when they are fed on:

1. Dry mature pastures, crops, straw, hay or maize silage, containing <0.7 mg α-tocopherol kg^{-1} DM (normal values for immature pasture are ⩾20).
2. Stored hay, rootcrops, crops or grain including acid-preserved high-moisture grain which may contain no vitamin E (Rammell, 1983).
3. Diets supplemented with polyunsaturated fatty acids or rancid fats, e.g. oilseeds and fish liver oil.

Dietary polyunsaturated fatty acids (PUFA) could be the precipitating cause of the infertility in cows fed on pasture grown on selenium-deficient soils and the response to selenium may depend upon the amount of PUFA, and the other anti-oxidants, α-tocopherol, β-carotene and possibly copper and molybdenum in the body.

Beta-carotene

Clinical β-carotene (and vitamin A) deficiencies occur when cattle are fed on diets containing <45 mg β-carotene kg^{-1} DM for prolonged periods.

Lotthammer and his coworkers (Ahlswede and Lotthammer, 1978; Lotthammer *et al.*, 1978) and others have suggested that β-carotene has a vital role in reproduction in cows that is unrelated to vitamin A. Goto *et al.* (1989) found that superovulated cows with high plasma β-carotene concentrations had higher embryonic survival rates (50%) than those with low concentrations (27%). Other work did not support Lotthammer's hypothesis (Folman *et al.*, 1979; Boyd *et al.*, 1984; Ducker *et al.*, 1984; Ascarelli *et al.*, 1985; Graves-Hoagland *et al.*, 1988; Wang *et al.*, 1988a,b). The latter two authors found no effect of β-carotene on plasma progesterone, corpora lutea size or progesterone concentration, frequency or amplitude of LH pulses, or release of LH in response to GnRH.

It is suggested that β-carotene may have been exerting an anti-oxidant effect, as it can do (Willson, 1983) in the absence of sufficient vitamin E/selenium. Maize silage, which formed the basic diet of the cows used by Lotthammer *et al.* (1978), is often deficient in selenium and vitamin E (Anon., 1983; Blood *et al.*, 1983; Putnam and Comben, 1987). The effect of β-carotene might be to protect the developing embryo.

An indication of the sensitivity of the ovum to mechanical injury was given by Legge and Sellens (1991) who were able to protect two-cell mouse ova during collection by adding reduced glutathione to the medium.

The anti-oxidants may also play some part in ruminal biotic metabolism. Hino *et al.* (1993) were able to prevent the depression in bacterial growth which was induced by the addition of sunflower oil.

Clinical Signs

Energy, protein and minerals involved in energy metabolism

The clinical signs shown by cows rendered infertile by diets deficient in energy, protein and the minerals involved in energy metabolism are as follows.

Signs referable to the reproductive system

These are, in order of increasing degrees of severity and duration of the deficiency:

1. Return to oestrus after unsuccessful mating or insemination, with variable cycle lengths (McClure, 1965a, 1970; Table 5.2).
2. Weakened intensity of oestrus (mild oestrus), shortened oestrus and suboestrus (Suzuki et al., 1982).
3. Delay in the recommencement of oestrous cycles post-partum (Gauthier et al., 1983; Butler and Smith, 1989).
4. Anoestrus.
5. Delayed puberty (Crichton et al., 1959; Sorensen et al., 1959; Reid et al., 1964).

Signs referable to systems other than the reproductive system

The clinical signs referable to the non-reproductive systems shown by cows affected by metabolic infertility, though usually invisible, can often be detected by measuring bodyweight and milk yields. Acute observation may reveal signs of central nervous system depression. The clinical signs shown by cows affected by (chronic) nutritional infertility (malnutrition) are usually visible and include subnormal growth rate, loss of weight and condition, reduced milk yield and quality (fat and protein) (Rook and Line, 1961; Foot et al., 1963; Storry and Rook, 1966), loss of elasticity of the skin, tendency to achromotrichia and central nervous system depression. Additional signs specific to aphosphorosis (lameness and pica), manganese deficiency (congenital deformity) and cobalt deficiency (pallid mucous membranes and conjunctiva) may be present.

Miscellaneous minerals and vitamins with a common anti-oxidant function

The clinical signs shown by the cows in herds with deficiencies in selenium, α-tocopherol, β-carotene or copper or which have appeared to respond to treatment with these substances have not been clearly defined. The evidence indicates that the signs are likely to be restricted to returns to oestrus after mating with cows returning after intervals of greater than 24 days.

5 DIAGNOSIS

Diagnosis of the specific causes of nutritional and metabolic infertility involves: (i) the recognition of reproductive failure; the identification of an infertility syndrome indicative of nutritional and metabolic cause and its differentiation from other causes of infertility, one or more of which may be coextant; (ii) finding evidence of the intake of insufficient or excessive amounts of nutrients essential for reproduction; abnormal concentrations of metabolites in the blood or other tissues; or the ingestion of poisons; and, finally, (iii) confirmation of the diagnosis by controlled therapeutic trials using adequately defined nutrients or metabolites.

Diagnosis is made difficult by the largely negative nature of the clinical signs shown by infertile cows; the delay between the onset of malnutrition or nutritional stress and the first appearance of clinical signs; the spontaneous recovery which may occur even before signs are shown; the similarity between the signs of reproductive failure caused by the various nutrient and metabolic factors, and between these and non-nutritional causes of infertility; and by the absence of satisfactory retrospective diagnostic tests. Clinical signs of nutritional and metabolic infertility are not pathognomic, either for nutritional and metabolic infertility in general or for its specific causes.

Notwithstanding its complexity, the diagnosis of nutritional and metabolic infertility follows the same principles as for the diagnosis of other diseases either in individual animals or herds. These are the consideration of history, clinical examination and the use of diagnostic aids or tests. Students will remember that of these three, clinical examination is the most important method of diagnosis of the cause of disease in individual animals, and that the history and results of laboratory or other diagnostic results are ancillary and indeed accepted only if they are consistent with the clinical findings. This is also true of herd problems, though history makes a greater contribution than it does for individual cases of disease.

The largely negative nature of the clinical signs of nutritional and metabolic infertility forces the clinician to place less reliance on clinical signs and more on history and diagnostic aids in establishing the diagnosis. Such techniques will, however, establish only tentative diagnoses. Definitive diagnosis must await the results of controlled therapeutic trials, in which a number of animals are treated with the suspected deficient nutrient or metabolite and others left untreated as controls.

Sequential trials, comparing the fertility of cattle after treatment with that existing before treatment, do not form a satisfactory basis for a definitive diagnosis.

History

Clinical

Useful histories of clinical signs include records of liveweight and condition together with records of signs suggestive of malnutrition (see p. 61) and those of clinical cases of other metabolic diseases.

Reproductive

The essential and desirable data on which the breeding history is based are described on p. 9.

If satisfactory breeding records are not available, it is often possible to build up useful records from the limited information which is available. Usually this includes identity, age, lactation number ($L = 1$, $L = >01$) and calving date. Pregnancy diagnosis by manual palpation *per rectum* or ultrasonography, accurate to within one oestrous cycle, provides reasonably accurate estimates of conception dates, calving-to-conception intervals and the proportion of the herd which will have inter-calving intervals of <400 days.

As explained on p. 8, routine monitoring of herd fertility identifies the existence of herd breeding problems. Further analysis of the data is necessary to define the infertility syndromes and reveal the probable pathogenesis. The diagnostician will, when analysing the records (either manually or by computer), plan the analysis so that all primary analyses can be done in the one pass. It may be necessary to make a secondary analysis based on the result of the first.

Analysis of breeding records

The analysis of the breeding records can be done simply by transferring the information about each cow to one or more cells of the forms shown in Table 5.1. This enables the determination of the following:

1. Parturition-to-first-oestrus interval and the frequency of occurrence of oestrus by parturition plus 60 days after calving.
2. Parturition-to-first-mating interval and the submission rate (proportion of the herd mated) by parturition plus 83 days.
3. First-service pregnancy (or non-return) rate.
4. Parturition-to-conception interval.
5. Pregnancy rate over the breeding season (or by three months after first service).
6. Abortion rate.
7. Perinatal mortality rate.
8. Cycle-length distribution.

relative to the:

1. Parturition-to-first-service interval.
2. Date of mating.
3. Mating method.
4. Identity of bull.
5. Age (or lactation number).

The herd infertility syndromes which may be revealed by analysis of the breeding records are:

1. Low pregnancy rate or non-return rate to first service.
2. Low proportion of cows pregnant by the end of the seasonal mating period or by three months after first service in non-seasonal mating herds.
3. Perinatal mortality.
4. Pre-weaning mortality.

Primary classification into one or more of these syndromes substantially reduces the range of diagnostic possibilities. Further reduction in the range of diagnostic possibilities can be achieved by comparing the reproductive performance of different breeding groups within the herd. Analysis of the records of Herd A (Table 1.4) are shown in Table 5.2.

Breeding groups are identified by selecting cows on the basis of:

1. Age and/or lactation number. The fertility of cows is highest soon after puberty and falls slowly for some years then rapidly from nine years of age (seventh lactation) (NZDB, 1961). Factors altering this potential fertility can include:

   (a) Delayed puberty caused by chronic malnutrition and chronic disease (e.g. parasitism).

Table 5.1. Analysis of breeding records. Part I.

(a) *First-service pregnancy/non-return rate.*[a]
First service >60 days after parturition/First service <60 days after parturition/Total[b]

Group		J	F	M	A	M	J	J	A	S	O	N	D	Total	J	F	M	A	M	J	J	A	S	O	N	D	Total
							19__													19__							
AI	L = ≥07																										
	L = 02–06																										
	L = 01																										
	Total																										
Natural service, Bull A	L = ≥07																										
	L = 02–06																										
	L = 01																										
	Total																										
Natural service, Bull B	L = ≥07																										
	L = 02–06																										
	L = 01																										
	Total																										
Total	L = ≥07																										
	L = 02–06																										
	L = 01																										
	Total																										

[a] No. of pregnant or not returned to oestrus/no. mated, e.g. 4/5.
[b] Delete two of the three headings – three sheets are required.

Table 5.1. Analysis of breeding records. Part II.

	L=01	L=02–06	L=≥07	Total
(b) *1st oestrus*				
$\bar{x}$ interval, parturition to 1st oestrus (day)				
% recorded in oestrus by parturition + 60 days				
(c) *1st mating*				
$\bar{x}$ interval, parturition to 1st mating (day)				
% mated by parturition + 83 days				
(d) *Conception*				
$\bar{x}$ interval between parturition and conception (day)				
Only AI				
AI + natural service				
Only natural service				
Total				
(e) *Pregnancy rate*				
% pregnant by 1st mating + 3 months				
Only AI				
AI + natural service				
Only natural service				
Total				

(f) *Abortion rate*
% aborting
  Only AI
  AI + natural service
  Only natural service
  Total

(g) *Perinatal mortality rate*
% deaths at/about parturition
  Only AI
  AI + natural service
  Only natural service
  Total

(h) *Cycle-length distribution*

	<18 days	18–24 days	>24 days
Last cycle before mating			
No.			
%			
1st cycle after unsuccessful first service			
AI			
No.			
%			
Natural service			
No.			
%			
Total			

Table 5.2. Summary of the reproductive performance of Herd A.

(a) Proportion mated in first four weeks of the breeding season.

	L=01	L=02–06	L=≥07	Total
No. mated	14	40	28	82
No. not mated	1	5	0	6
% mated (target 100%; problem <90%)	93	89	100	93

(b) First-service non-return rate.

		>60 days post-partum		<60 days post-partum		Total	
		Nos.	%	Nos.	%	Nos.	%
AI	L=≥07	5/17	29	2/10	20	7/27	26
	L=02–06	3/7	43	0/0	–	3/7	43
	L=01	5/6	83	0/1	–	5/7	71
	Total	13/30	43	2/11	18	15/41	37
Natural service	L=≥07	1/1	–	0/0	–	1/1	–
	L=02–06	10/33	30	1/5	20	11/38	29
	L=01	4/7	57	0/1	–	4/8	50
	Total	15/41	37	1/6	17	16/47	34
Total (target >75%; problem <60%)		28/71	38	3/17	18	31/88	35

(c) Proportion not pregnant at the end of the breeding season.

	L=01	L=02–06	L=≥07	Total
Only AI	0/7	1/6	2/18	3/31
AI+natural service	0/0	1/1	2/9	3/10
Only natural service	1/8	4/38	0/1	5/47
Total (target ≤2%; problem ≥10%)	1/15 (7%)	6/45 (13%)	4/28 (14%)	11/88 (13%)

(d) Cycle-length distribution.

		1st return cycle, no. and (%)			
		<18 days	18–24 days	>24 days	Total
AI	L=≥07	3	11	6	20
	L=02–06	1	2	1	4
	L=01	0	2	0	2
	Total	4 (15%)	15 (58%)	7 (27%)	26
Natural service	L=≥07	0	0	0	0
	L=02–06	12	10	5	27
	L=01	2	2	0	4
	Total	13 (42%)	12 (39%)	6 (19%)	31
Total		17 (30%)	27 (47%)	13 (23%)	57
Target		≤5%	85%	≤15%	
Problem		≥10%	55%	≥35%	

(b) Lactation anoestrus or suboestrus and low first-service pregnancy rates in first-lactation heifers, caused by malnutrition during late pregnancy whilst the heifers are still growing, and continued on through early lactation.

(c) Low first-service pregnancy rates in cows which had been well fed up to calving but underfed during the early part of lactation up to and including the mating period. If all the cows in the herd are offered the same feed, e.g. pasture (grazing) without selective supplementation, the younger cows, i.e. the first-lactation cows, produce less milk and tend to be more fertile than adult cows.

(d) Infertility in first-lactation cows with a history of uterine disease during the first pregnancy or at about the time of first parturition (e.g. brucellosis and dystocia).

(e) Infertility in first-lactation cows in herds with endemic venereal infections where the practice is to mate maiden heifers to young bulls separately from the adult herds and which have been exposed to the infections from the herd bull for the first time. These are relatively infertile compared with the older cows which have developed some degree of immunity. In recently infected herds, cows are not immune and all are equally affected.

2. Mating method. Comparison of the fertility of cows naturally served with those artificially inseminated provides a valuable clue as to whether the infertility is of male or female origin, and if male whether it is due to the mating method. Care must be taken to ensure that the choice of mating method was made without bias. Cows affected with nutritional and metabolic infertility generally perform poorly whether they are naturally mated or artificially inseminated.

Infertility restricted to the naturally mated cows indicates male infertility or venereal infection. Infertility restricted to the artificially inseminated cows indicates faulty heat detection or AI technique (including the use of poor-quality semen).

3. Identity of the bull or source of semen. Comparison of the fertility of cows inseminated with semen from different bulls and different ejaculates from the same bull and cows naturally mated by different bulls helps to identify infertile bulls and bulls carrying venereal infections or semen of poor quality.

4. Parturition-to-first-service interval. The fertility of cows at the time of parturition is zero. Fertility returns to normal for physiological reasons on average at about 60 days post-partum (VanDemark and Salisbury, 1950). In the consideration of any apparent herd infertility problem, it is necessary to exclude those cows which are infertile for physiological reasons. This can be done by grouping the herd into two groups, those mated for the first time <60 days and those mated for the first time >60 days post-partum (Table 5.1).

5. Lactation. Nutritional and metabolic factors must be considered when an association is found between lactation and fertility, e.g. when the fertility of the cows lactating during the mating season is lower than the fertility of the non-lactating cows. In most seasonally mated herds there will be a few (2–3%) cows which did not become pregnant the previous season and are now either dry or producing little milk ('carry-over cows').

6. Liveweight, liveweight change and condition. In general, liveweight, liveweight change and condition are indicators of the state of nutrition of the herd or groups of cows within the herd. If reproductive failure is found to be related to weight or condition, it is likely, though not certain, that the cause is nutritional or metabolic. Infectious and parasitic diseases of the herd need to be considered in the differential diagnosis.

7. Milk yield and composition. Although the evidence for the reported correlation between high milk yield and infertility having a nutritional basis is somewhat tenuous, it is worthwhile comparing the fertility of high and low yielding cows, measured at about the time of mating.

8. Mating date. Chronological analysis of mating records may reveal seasonal variations in fertility. Significant variations in fertility are useful indicators which, when correlated with other data, e.g. introduction of a new bull or change in diet or dairy staff, help with the diagnosis. Nutritional and metabolic factors may cause changes in fertility which can be correlated with preceding changes in the diet. It should not be assumed, however, that a reduced fertility following a change in diet is necessarily caused by the new diet.

Chronological analysis is of only limited value in seasonally mated herds, particularly when the mating season is short, and can be misleading when the post-partum interval is ignored.

9. Other criteria. Preliminary investigations of a herd infertility problem are likely to suggest additional criteria for selecting cows for breeding group comparisons. These include the identities of inseminators and dairy staff who detect oestrus, breed, source of stock and date of entry into the herd, and recent vaccination history. It may be necessary to make a subsidiary analysis of the breeding records to elucidate finer points. Flexible programs are necessary if computers are to be used for the subsidiary analyses.

The cycle-length distribution of the cows before mating and those returning to oestrus after unsuccessful first services should then be determined for each breeding group.

Interpretation

Nutritional and metabolic infertility are indicated by: delayed puberty or first post-partum oestrus; abnormal oestrus; low first- and subsequent-service pregnancy rate; abnormal oestrous cycle lengths; or abortion, congenital deformity and perinatal mortality, that is/are associated with: age and lactation number; season and pasture growth or change in feeding practice; lactation, milk yield and composition; suckling; liveweight/condition or changes in liveweight and condition; season of the year; nutrition (feed type and quality including batch and soil type); other metabolic diseases, particularly ketosis; but not with mating method (AI, natural service or specific bulls).

Some such associations may also occur in the presence of other causes of reproductive failure.

Abnormal manifestation of oestrus must be differentiated from failure to observe or record oestrus. This is indicated by breeding records which show long intervals from calving to first oestrus and inter-oestrous intervals of n (18 to 24 days) unrelated to breeding group but related to staff on heat-detection duty in hand-mated and artificially inseminated herds.

There is an alternative explanation for some of the herd infertility problems of the low first-service pregnancy (or non-return) rate type, viz., erroneous heat detection, identification or recording. In these cases there will be correlation only with hand service or AI and with staff on duty, and little or no correlation in other breeding groups. There should also be large secondary and tertiary peak frequencies at 2(18–24 days) and 3(18–24 days).

The cycle-length distributions before and after unsuccessful first service reflect the pathogenesis and assist in determining the cause of reproductive failure. Three distinct patterns occur.

1. A normal cycle-length distribution (5% of cycles <18 days, 85% between 18 and 24 days and 10% >24 days) before mating indicates normal hypothalamic, pituitary and ovarian function. Returns to oestrus after unsuccessful first service within these limits indicate fertilization failure or death of the ova before D16.

2. Cycle lengths which are more widely distributed about the mean occur in groups of cattle affected with hypothalamo–pituitary–ovarian failure including delayed recurrence of post-partum oestrus. For example, in Herd A (Table 5.2) 30% of the cows' cycle lengths were <18 days and 23% >24 days.

3. When cows return to oestrus after mating with lengthened cycles, there seem to be only two or possibly three possible pathogeneses. Most likely is the possibility of embryonic mortality after D15, in which case there will be abnormal lengths only after mating. Also possible is the development of suboestrus in cows which fail to become or remain pregnant after service. It is likely in this case that there will also be some abnormality of pre-mating cycles. The third possibility is that errors have been made in heat detection.

When comparing pre- and post-mating cycle lengths, due regard needs to be paid to the fact that the first post-partum oestrous cycle may be abbreviated.

Examples

1. Analysis of the breeding records of a seasonally mated pasture-fed herd (Herd A, Table 1.4) supplying milk for manufacture is shown in Table 5.2. In this herd it is necessary that the cows calve as near as possible to the required date in order to superimpose the nutritional demands upon the seasonal pasture production (Fig. 1.1).

Analysis of the breeding records shows that this seasonally mated herd experienced an infertility problem characterized by a low first-service non-return rate which failed to recover spontaneously during the breeding season causing cows to calve late, or not in time for the next calving season. Comparison of the breeding groups indicates that the problem was female in origin and that the older cows were affected to a greater degree than the young ($L = 01$) cows. Cows returned to oestrus after unsuccessful first services at irregular intervals with more short and more long cycles than normal, indicating that the pathogenesis involved ovarian dysfunction. A retrospective tentative diagnosis of hypoglycaemic infertility was made in this case.

2. Analysis of the breeding records of a non-seasonally mated dairy herd (Herd B, Tables 5.3 and 5.4) affected with chronic selenium-deficiency- (? 'selenium-responsive') induced infertility. Over two years, this herd had a low first-service non-return rate, long calving-to-conception intervals and lengthened return-cycle intervals, suggestive of late embryonic mortality.

Differential diagnosis

The systematic analysis of the breeding records in this manner also assists in differential diagnosis. This can be illustrated by the analysis of the records (Table 5.5) of an infertile spring-mated dairy research herd in which both AI and natural service had been used. The problem was characterized by a low first-service non-return rate with failure to recover spontaneously during the breeding season, i.e. a problem similar to Herd A (Table 5.2). Campylobacteriosis (vibriosis) was diagnosed and confirmed bacteriologically and the cows were treated in the fifth week of the breeding season without any apparent response. Retrospective analysis of the records showed that the *Campylobacter fetus* infection had been introduced into a herd with basic female infertility problem of a type shown in the analysis of Herd A.

Nutritional

An accurate nutritional history can be obtained only when the intake has been measured and samples of the feed analysed (see p. 92).

In the absence of records of feed intake and composition, some indication of the nutritional history can be obtained by close questioning of the

Table 5.3. Herd breeding records of Herd B.

HERD: B ADDRESS: DEPT. VETERINARY CLINICAL STUDIES, THE UNIVERSITY OF SYDNEY

NAME OR NO.	Age	Calving Date	Details	1981												OESTRUS AND MATING DATES AND IDENTITY OF BULL						Calving Date	Calving Details	1982											
				J	F	M	A	M	J	J	A	S	O	N	D									J	F	M	A	M	J	J	A	S	O	N	D
340	12													30A 10A	P 20A	8/4	RFM											22A	28A	19A	10A		
351	12	29/8														21/8	RFM																
389	12	21/7	RFM																										SOLD				
399	11															22/8														22	3		
414	11								27A 4	13A	30A 8A	P 19A			13/4							25			16A	4AP					
426	11	12/7														28/9								28A 4	19AP								
428	10															2/3																	
441	10															28/2						21	28A 5A	3A 23A	3A 17A	6AP 27A 3A 15A	25A					
449	10	4/10	ENDO									23	23				4A	23	19A							13A				
453	10											5.19			30/5	RFM																	
480	10	31/7						29				4A	15AF						20	20A 2A	11	2A	2A	13A	2A			
482	9															5/2	ENDO		20	1	30A 5A	23A	17AP						
603	9															9/3					9 23A 2A	2A 22	24AP 5A						
618	9	7/12	MF									18	23AP				15	25	23			1AP					
624	9							24A 2A	16A			10A	2A	5A	1A		19/6	RFM												
625	9	7/2	RFM	2											5/9	MF												
628	9															3/8													24A		
630	8						P Abortion	15A 8.29	20A	28AP			21/3						19AP						
659	8	2/2		20		20A 1									3/9														
666	8	20/10															15	8AP												
667	8															24/4							20	15	9A	17AP		
672	8															8/2				7A	25A 3A		26A 5A					

HERD: B **ADDRESS:**

DEPT. VETERINARY CLINICAL STUDIES, THE UNIVERSITY OF SYDNEY

OESTRUS AND MATING DATES AND IDENTITY OF BULL

Name or No.	Age	Calving Date	Details	1981 J	F	M	A	M	J	J	A	S	O	N	D	Calving Date	1982 J	F	M	A	M	J	J	A	S	O	N	D
676	8	18/12																11A	3AP									
677	8																			25				27			
682	8	24/11	RFM															17.24/11	15AP									
686	8	15/10											15	28AP/4		5/10								8		18		
687	7	27/8																										
689	7	8/4																										
691	7	16/1	RFM			21	13A													27	20AP							
693	7	10/7	RFM						15AP		28						TO BE CULLED										
700	7															20/2	12A	4A.10/14.24	25A	13A	23A	5A 31AP / 9A 11N 27A	12/22	10A	20AP 30	SOLD		
702	7	28/8									30		21A	10A/22A														
705	7											6	25	18A/30A		23/5	11A	9N/27A	8	16A		21A 23A/20A/24A		15AP	1A 25A			
708	7	9/8														1/1	12.20	16AP										
711	6															29/6												
712	6															5/3				19	9N 31AP							
713	6															29/7								29	22	13		
715	6															18/5						30		15AP				
725	6															31/1												
730	6											5/30	21A	7A			23A	18 9A/14	17AP	26	23A	16A 6A/31A/4		20A		6		
740	6	22/8																	2A/25	17A	SOLD	30A	17Ap					
744	6															20/5	15	7										
745	6	20/12																8A/27A 16				21A 11AP						
746	6	3/11											25	15														

Table 5.3. (continued).

HERD: B ADDRESS: DEPT. VETERINARY CLINICAL STUDIES, THE UNIVERSITY OF SYDNEY

NAME OR NO.	Age	Calving Date	Details	1981												Calving Date	Calving Details	1982											
				J	F	M	A	M	J	J	A	S	O	N	D			J	F	M	A	M	J	J	A	S	O	N	D
757	6															19/1					21		27	16A	13AP				
767	6															11/3	RFM						8	21A 12A	2A 23A		26		
768	5															2/8	DYS					17	31A	DIED					
769	5															21/2					6	7 19A 5A	10 27	11A	20A	9AP 5A			
770	5															4/2			16	15 6	26A 16N 20	23AP	27A 19A			27A			
771	5															1/2				28A	20A 30A								
775	5															31/7					10					30			
780	5								13			8A 29A				28/2					27	23AP							
783	5	14/6	ABORT																2A 23A 14A		2A 23A 15NP								
785	5															24/3	RFM				10	1	20A		26AP				
786	5															24/1				SOLD									
788	4															2/3		4AP			5AP								
789	4	27/8												5	15A	12/10													
790	4															24/1		12	27A 21A	20A	26AP 2A 20AP				2A 21AP				
791	4	27/11													7 26A			8A		19A	3		14A 30A 19A		20A	15A 11A			
792	4	9/9												25															
793	4															10/8			16	8	11A 14N		7AP		20				
795	4															4/2		17	27A		7AP								
796	4	25/12																	17AP										
797	4	24/12																											
798	4													1	28AP	21/1		27	20A	10AP									
799	4	21/8														10/9													

HERD: B   ADDRESS:

DEPT. VETERINARY CLINICAL STUDIES, THE UNIVERSITY OF SYDNEY

OESTRUS AND MATING DATES AND IDENTITY OF BULL

NAME OR NO.	Age	Calving Date	Details	1981 J	F	M	A	M	J	J	A	S	O	N	D	Calving Date	Details	1982 J	F	M	A	M	J	J	A	S	O	N	D
801	4	6/9												2 23A P		5/9				25	22A		11A P				26			
802	4																7/1									9A 31A		20A	12A			
803	4																10/4							27							
806	4	6/9	RFM									18		25 11.17	19A P	26/9	RFM													
807	3																29/3							29	18	29A	22A		25A			
808	3	30/9												1 24	13A					13A 27	22A P									
809	3	21/10												15	14					4A 27A	17A P										
815	3	18/12												11	28A P					15	28A P										
817	3	20/8												30A																	
820	3	21/8											25	28A	22A	5/9					2A 23A		29A	25A	31A			13A		
821	3	11/8											8	19A	30A P	17/9															
822	3																30/7													13A		
824	3	15/9												18A	31A	31A				18A	5 26A			5A		2.24A 7 30 13A			SOLD		
825	3	12/9											18	17	13A P										27A				22A		
826	3																4/8								2A 23A 16N	2A	20	4 24			
827	3	15/12																		23		9A	7 27	19A		11A	17A	9A	4A		
830	3																10/3				5 28A			3A P			23A P				
831	3	6/12	RFM														7/3	RFM					19A			22A 30A	21A 28		6A		
832	3																						5A P	30 27							
833	2	20/12																		23	11			20 28	11.1B 2 23 24A	16A	2 24A	7A P			
835	2																10/5							27	21A	28A	20A	9A			
836	2																22/3															

Table 5.3. (continued).

HERD: B   ADDRESS: DEPT. VETERINARY CLINICAL STUDIES, THE UNIVERSITY OF SYDNEY

NAME OR NO.	Age	Calving Date	Details	\multicolumn{12}{c	}{1981 — OESTRUS AND MATING DATES AND IDENTITY OF BULL}	Calving Date	Details	\multicolumn{12}{c	}{1982}
				J	F	M	A	M	J	J	A	S	O	N	D			J	F	M	A	M	J	J	A	S	O	N	D
837	2	11/8												11 / 30A	19A			8A / 27A 12		2A / 21N	11A / 3	21A	11A / 29A / 5P	18A / P	8P		18A		
838	2																				14	23A 16		24	16A				
840	2															12/2						30	27	15AP					
841	2															20/6	RFM DYS					17	26A	14A	3A	22A			
843	2															21/3					27					13A				
845	2															5/3													
846	2	3/7	RFM							7 / 30A			23A		15/8		2AP			29		7	30A	20AP				
847	2	1/12													13/10		2	17A					28	17	8AP				
849	2															3/6								4		23AP				
851	2															22/5							25		3A	18A	7A			
853	2															2/5	DYS												
855	2															13/6	RFM						25	15A	27A				
859	2															5/5					20		27A	17A	9A / 29A				
861	2															22/4					23	13		21A	12A				
862	2															1/7								8		2A			
864	2															27/4							18A			10			
867	2															12/4					8 / 29	17A	9AP						
868	2															10/5							27AP						

Ages: 2 years = L = 01; 3–7 years = L = 02–06; 8–12 years = L = ≥07. Details: DYS, dystocia; ENDO, endometritis; MF, milk fever; RFM, retained fetal membranes. Mating date codes: A = AI; N = natural service; P = pregnant at diagnosis.

Table 5.4. Analysis of herd breeding records of Herd B.

(a) First-service pregnancy rates.

Parturition-to-first-service interval	Lactation no.	1981													
		J	F	M	A	M	J	J	A	S	O	N	D	Total
>60 days	L=≥0.7										0/1	0/1	1/1	2/6
	L=02–06					0/1					0/2	2/7	3/6	5/16
	L=01									0/1		0/1		0/2
	Total					0/1				0/2	0/3	2/9	4/7	7/24
<60 days	L=≥07								0/1		0/2			0/3
	L=02–06			0/1	0/1		1/1							
	L=01								0/1		0/2			0/3
	Total			0/1	0/1		1/1							
Total	L=≥07					0/1			0/1		0/1	0/1	1/1	2/6
	L=02–06			0/1	0/1		1/1			0/1	0/4	2/7	3/6	5/19
	L=01								0/1	0/1		0/1		0/2
	Total			0/1	0/1		1/1		0/1	0/2	0/5	2/9	4/7	7/27

Parturition-to-first-service interval	Lactation no.	1982														1981 and 1982	
		J	F	M	A	M	J	J	A	S	O	N	D	Total	Total	%
>60 days	L=≥07			0/1	1/1	1/1	0/2	0/2	0/1	0/2	0/1		0/1	2/12	4/18	22.2
	L=02–06			0/2	1/4	2/5	1/4	2/6	1/6	0/3	1/1	1/2		9/33	14/49	28.6
	L=01					1/1	0/2	0/2	1/2	1/5	2/4		0/1	5/17	5/19	26.3
	Total			0/3	2/5	4/7	1/8	2/10	2/9	1/10	3/6	1/2	0/2	16/62	23/86	26.7
<60 days	L=≥07				0/1			0/2						0/3	0/3	
	L=02–06				0/2	0/2	1/1				1/1	0/1		1/5	1/8	12.5
	L=01						1/1				1/1			2/2	2/2	
	Total				1/3	0/2	1/1	0/2				0/1		3/10	3/13	23.1
Total	L=≥07			0/1	1/2	1/1	0/2	0/4	0/1	0/2	0/1		0/1	2/15	4/21	19.0
	L=02–06			0/2	2/6	2/7	1/4	2/6	1/6	0/3	1/1	1/3		10/38	15/57	26.4
	L=01					1/1	1/3	0/2	1/2	1/5	3/5		0/1	7/19	7/21	33.3
	Total			0/3	3/8	4/9	2/9	2/12	2/9	1/10	4/7	1/3	0/2	19/72	26/99	26.3

Table 5.4. (continued).

(b) Other measures (1981, 1982).

Measure	Lactation number			
	L=01	L=02-06	L=≥07	Total
Mean interval (days):				
Parturition to first oestrus	36.85±21.0	35.37±20.44	45.91±23.54	38.62±21.14
Parturition to first mating	79.55	76.16	74.16	76.58
Parturition to conception	135.40	149.27	131.71	142.86
No. and proportion:				
Recorded in oestrus by parturition+60 days	15/22 (68.2%)	46/66 (69.7%)	16/31 (57.6%)	79/120 (65.8%)
Mated by parturition+83 days	14/22 (63.6%)	37/60 (61.9%)	15/30 (50%)	64/112 (57.1%)
Pregnant by first service+3 months	9/21 (42.9%)	30/43 (69.7%)	15/23 (65.2%)	54/97 (56.7%)
Aborting	0/22 (0%)	2/64 (3.13%)	1/31 (3.23%)	3/117 (2.56%)
Calves dying at/about birth	0/22 (0%)	0/64 (0%)	0/31 (0%)	0/117 (0%)

	Cycle-length distribution (days)											
	<18	18-24	>24	<18	18-24	>24	<18	18-24	>24	<18	18-24	>24
Last premating cycle n	0	11	7	0	29	18	1	8	6	1	48	31
%	0	61	39	0	62	38	7	53	40	1	60	39
1st return cycle after unsuccessful mating n	0	7	7	0	17	21	1	7	8	1	31	36
%	0	50	50	0	45	55	6	44	50	1	46	53

Table 5.5. Analysis of breeding records of a herd with both vibrionic (campylobacteriosis) and metabolic infertility.[a]

Breeding group	First-service non-return rate, cows mated for the first time				
	>60 days after parturition		<60 days after parturition		Total
Artificially inseminated cows					
L=≥02	6/16	38%	1/2	–	7/18 39%
L=01	1/5	20%	0/1	–	1/6 17%
Total	7/21	33%	1/3	–	8/24 33%
Naturally mated cows					
L=≥02	1/10	10%	0/0	–	1/10 10%
L=01	3/5	60%	0/1	–	3/6 50%
Total	4/15	27%	0/1	–	4/16 25%
Total					
L=≥02	7/26	27%	1/2	–	8/28 29%
L=01	4/10	40%	0/2	–	4/12 33%
Total	11/36	31%	1/4	–	12/40 30%

[a] Analysis of the data collected in the course of research showed that the first-service non-return rate of the cows which were gaining weight at the time of mating was 59% compared with 14% for those losing weight, irrespective of their *Campylobacter fetus* antibody status.

manager/dairyman as to the type and quantities of roughage and concentrates fed, analyses of manufactured feedstuffs purchased, and in the case of grazing cattle, pasture/crop type, its length, density and stage of maturity, and access the cows had to the feeds. Pasture/crop fertilization and soil composition data may also be available.

At this stage, after analysis of the breeding records and consideration of the nutritional history it is often, perhaps usually, possible to make a tentative diagnosis of nutritional or metabolic infertility. It may be possible to find support for such a general or even specific diagnosis by clinically examining the cattle, clinical pathology and feed analysis. It is also possible and even probable that the nutritional and metabolic factors are no longer operating at the time of clinical examination and testing. Examinations are best carried out at the critical stages of reproduction, immediately before mating heifers, at calving and five to eight weeks after calving (just before mating), to diagnose the cause of potential infertility for the purposes of control and prevention.

Clinical Examination

Cows are selected from the class of cattle indicated by the analysis of the breeding records to be at risk, and examined for signs of malnutrition. The general clinical examination should include liveweight, liveweight change and condition, and evidence of abrasions over the sciatic tubers and sacrum. The internal genitalia are examined manually *per rectum* and/or ultrasonically taking particular note of the state of ovarian activity and fetal development.

At least 20 cows of the group at risk should be examined at seven-to-ten-day intervals until the clinician is satisfied that the pathogenesis has been determined.

Measurement of liveweight and liveweight change requires the use of scales, and measurements being made at least twice, once at the time of mating and once one to two weeks previously. Additional measurements at two months and just before parturition, and at the time heifer calves are joined, are desirable.

Being independent of rumen-fill and stage of pregnancy, and relatively unaffected by body size, body condition provides a valuable indication of nutritional state. The condition is scored by visual estimation and palpation of muscle development and subcutaneous fat thickness over the lumber vertebrae half-way between the iliac tuber and last rib, and the fat round the tail head. The condition is scored from very thin to very fat in eight to ten grades (including half scores in the 1–5 scale) according to the system used (Earle, 1976; Kilkenny, 1978; Scott and Smeaton, 1980).

Clinical Pathology

Clinical pathology tests can be performed on blood and milk samples taken from groups of seven to ten cows selected on the results of analysis of breeding records. Single samples are satisfactory where chronic mineral deficiencies are suspected; two samples are preferred from cattle with metabolic infertility taken one or two weeks before and at the time of mating. This is to detect changes in values and to measure the plasma progesterone concentration when it should be at its peak during the dioestrus before mating. Where possible, samples should be taken from seven to ten cows likely to be shown on analysis of the records to be infertile and a similar number from a group of cows at the same stages of reproduction or lactation and which are likely to be fertile (e.g. on the basis of age or milk yield). The mean values of the two groups are compared with each other and the normal values for the classes of cattle. Samples taken eight to ten weeks after calving are likely to yield the most useful results.

The tests to be performed will depend upon the tentative diagnosis made after considering the history and clinical signs shown by the infertile cows. Those likely to be of value in the field include:

Milk: Milk fat, protein, progesterone (oestrus minus 8–15 days).
Blood: Progesterone (oestrus minus 8–15 days; Folman et al., 1973). Where energy balance appears to be involved glucose, β-hydroxybutyrate and phosphorus.
Also, where early embryonic mortality is suspected selenium, α-tocopherol, β-carotene and copper.

Two factors mitigate against the chances of identifying nutritional and metabolic imbalances by clinical pathological tests: (i) the causes may no longer be operating at the time of sampling; and (ii) normal homeostatic mechanisms attempt to keep blood concentrations within normal limits.

Differential Diagnosis

It is not usually possible either to differentiate nutritional and metabolic infertility from other causes of reproductive failure or to differentiate between various nutritional and metabolic causes on clinical grounds alone. The most powerful diagnostic tools for this are the comparison of the fertility of the various breeding groups within the herd, the analysis of the cycle-length distributions and the use of controlled treatment trials.

Controlled Therapeutic Trials

Confirmation of tentative diagnoses of nutritional and metabolic infertility is possible only by conducting controlled field trials in which the diet of a sample of the cows is amended to overcome the suspected deficiency or excess. The trials should follow the standard methods for field trials, paying particular attention to randomization, selection of susceptible animals and comparable control animals, and the use of sufficient numbers to enable the results to be statistically analysed.

In the planning of therapeutic trials for the confirmation of tentative diagnoses of nutritional and metabolic infertility, cattle should be paired on the basis of age and lactation number, calving date, calving history, stage of lactation (post-partum interval), yield and breed, all factors which affect reproductive performance. The trials can be described as being stratified randomized.

Factors which should be kept constant are: nutrition in respect to all nutrients other than that being tested, mating method, bull/semen batch and heat detection (AI). Most parameters can be analysed and the results compared using student's t test and the chi-square test. The numbers of animals available in a herd, particularly one with all-year-round mating, may be too small to enable such tests to achieve statistical significance. One then has to be content with trends, and at times make recommendations for control measures on data which may not be statistically significant.

Field trials to test hypotheses of nutritional and metabolic causes of infertility are difficult in commercial herds for they are disruptive to herd management. The administration to individual cows of copper and selenium which have a relatively long-term effect (>4 weeks) presents no problem. Administration of nutrients that have effect for only hours is simple in hand-fed cows which are individually fed, but difficult in herds in which feeding is not individual, e.g. pasture-fed herds.

Therapeutic trials for the control of diseases where macronutrient deficiencies are suspected but the real causes are micronutrient or trace-element deficiencies are especially hazardous. In such trials undetected or ignored micronutrients and trace elements may be present in the feedstuffs designed or chosen to supply the macronutrients.

The trials that can be conducted are those which are described in Chapter 6, but limited to a sample of the herd. The number of animals in the treatment and control groups necessary to achieve statistical significance will depend upon the changes in fertility expected. Examples are shown in Table 5.6.

Owners and managers usually prefer to modify the diets of affected cows and herds on the basis of tentative diagnosis rather than await the results of controlled therapeutic trials.

Table 5.6. Examples of the numbers of animals required in each of treatment and control groups necessary to obtain significance at the 5% level.

Parameter	Control group	Treatment group	Test	Approximate minimal no. cows in each group	Value
First-service pregnancy or non-return rate	32%	60%	χ^2	25	$\chi^2 = 3.95$
Service : conception ratio	2.0 : 1 SD 0.8	1.5 : 1 SD 0.4	t^a	17	$t = 2.25$
% not showing oestrus by 60 days after calving	30%	9%	χ^2	33	$\chi^2 = 4.69$
Calving-to-first-oestrus interval (days)	$\bar{x} = 100$ SD 60	$\bar{x} = 50$ SD 20	t^a	11	$t = 2.50$
Calving-to-conception interval	$\bar{x} = 118$ SD 60	$\bar{x} = 83$ SD 10	t^a	20	$t = 2.50$

[a] t test number based on $n = \dfrac{t^2(SD_1^2 + SD_2^2)}{(\bar{x}_1 - \bar{x}_2)^2} + 1$.

TREATMENT, CONTROL AND PREVENTION

Nutritional and metabolic infertility are treated, controlled and prevented by feeding the cows according to their requirements for maintenance, exercise, growth, pregnancy and lactation, as described in Chapter 2. Any modifications to the diet necessary to treat infertile cows and control the infertility in infertile herds is urgent and therefore potentially more expensive than those required for prevention. Most cows and herds respond to the appropriate changes in the diet within three to six weeks (Wiltbank et al., 1964; Girou and Brochart, 1970a,b; Dindorkar et al., 1982) or possibly a little longer (nine weeks, Richards et al., 1989a,b).

Some aspects of treatment, control and prevention need special mention: the composition of the roughage and concentrate components of the diet; the very high-producing cow; monitoring nutrition, clinical signs and clinical pathology; heat detection; hormone therapy and weaning.

Roughage

When fed as the sole diet for lactating dairy cows, pastures and crops grazed, zero-grazed or conserved and hand-fed, need to be fed or conserved and fed at the optimal stage of growth, i.e. just before the flower stems emerge, when the yield of digestible organic matter (DOM) is at its peak and in unlimited quantities. At least 1500 kg dry matter (DM) ha^{-1} should remain after grazing. Methods for estimation of this are given on p. 93.

Well-fed Jersey cows of high genetic merit have yielded the quantities of milk shown in Table 6.1 from good quality pasture whilst maintaining high fertility (McClure, 1966). On a pro-rata basis Holstein-Friesian cows of 500 kg bodyweight can be expected to yield 6000 kg fat-corrected milk

Table 6.1. Milk production standards for grazing Jersey cows.[a]

Age (years)	No. of cows	Milk (kg)	FCM (kg)	Fat (%)	Fat (kg)
2	7	3109	3893	5.68	177
3	7	3155	4289	6.39	202
4	28	3765	4755	5.75	216
Total	42	3554	4533	5.84	207

[a] Controlled grazing, light stocking rate group, Ruakura Animal Research Station, 1959–1960. Mean seasonal liveweight: spring 367 kg, summer 388 kg, autumn 400 kg (McMeekan and Walshe, 1963).

(FCM) per lactation, a figure which is consistent with the theoretical yields based on metabolizable energy (ME) intake of merit cows fed unlimited amounts of good quality pasture.

If feed is short and cattle must be underfed, cows can be safely underfed between the time of implantation and the end of the seventh month of pregnancy without effect on their fertility provided that they are well-fed subsequently.

Very immature crops and grass-dominant pastures can be fed if supplemented by green legumes, hay, silage or preferably grain-based concentrates containing a reasonable rumen-undegradable protein (UDP) component. These allow a higher intake of DOM and, in particular, readily fermentable carbohydrate and protein yielding more volatile fatty acids and non-ammonia nitrogen to enter the duodenum (Warnick et al., 1955; McClure, 1965a,b, 1970; Parker, 1966; Ulyatt, 1969; Rogers et al., 1982; Thomson et al., 1985).

If sufficient pasture is not available it needs to be supplemented with the most economical balanced feedstuff procurable (counting the cost of freight, food and labour).

Where sufficient feed is available but its composition is such that the diet does not provide sufficient of each of the essential nutrients to meet the animals' requirements, the diet needs to be supplemented with the appropriate nutrients. These will vary according to the composition of the pasture.

Unlike very immature grass which is deficient mainly in readily available carbohydrate DM and UDP, the vegetative parts of grasses and crops at the flowering and later stages of growth contain an increasing proportion of fibre and a decreasing proportion of available carbohydrate, protein, minerals and ultimately carotene. Decreasing digestibility also reduces gross intake and reduces the intake of essential nutrients gradually up to the time of seed-fall, thence rapidly. It becomes increasingly difficult to balance such maturing grasses and crops. Initially, attempts should be made to maintain its digestibility by supplementing the feed with highly digestible young plants or, failing this, protein-rich supplements such as cottonseed meal which slowly releases nitrogen (N) and energy into the rumen (Foster et al., 1945; Warnick et al., 1955; Norman, 1963; Siebert et al., 1975; Liebholz and Kellaway, 1984) or a combination of protein and non-protein nitrogen (NPN) as described by Donaldson (1966). In the longer term it may be possible to improve the quality of the diet by improving the pastures by incorporating clovers and temperate-climate grasses or with tropical legumes (Holroyd et al., 1963; Norman, 1963).

Fibrous roughages, though low in N, are only marginally deficient in N because of their low potential digestibility (Orskov, 1982). Therefore, small responses only can be expected from the addition of NPN. For example, by feeding 69 g urea day^{-1} together with 0.21 kg molasses day^{-1} Winks et al. (1972) increased liveweight changes only from -0.01 kg day^{-1} to 0.23 kg day^{-1} and Campling et al. (1962) increased the digestibility of straw from 39.3 to 47.3% by intra-ruminal infusion of 150 g urea day^{-1}.

Ultimately it becomes impossible to balance the diet of cows fed indigestible grasses and it becomes necessary to restrict the access of the cows to such pastures and feed alternative feeds. The critical stage of maturity of the grass or crop varies with the nutritional requirements of the cows, the stage being related to milk yield and stage of pregnancy.

Pastures and crops grown on mineral-deficient soils may be deficient in the minerals essential for reproduction. Once accurate diagnoses have been made, and in the current state of knowledge these can be made only after conducting controlled treatment trials, the deficient minerals can be added to the soils as fertilizers, fed as supplements or administered directly to individual cows. In the first instance one of the two latter methods should also be used to obviate the delay which is unavoidable with pasture fertilization. The method used will depend upon cost, convenience, persistence of effect and the plant's requirements for the mineral.

Where the pasture productivity is high or is potentially so, the minerals required in large quantities such as phosphorus, and those which are required frequently such as cobalt (Co), manganese (Mn), phosphorus (P), and to a lesser degree, selenium and copper are best added to fertilizers and applied to the soil. Where the pasture productivity is low and the application of fertilizers uneconomic, the minerals can be added to feed supplements or to licks (iodine, phosphorus) or incorporated in slow-release capsules or pellets administered per os (cobalt, copper and selenium) or injected parentally (copper and selenium). Copper and phosphorus, being essential for plant growth, are best applied as fertilizers where the pasture production is sufficient to make this method economic. The trace elements required by animals but not by plants (cobalt and selenium) may be administered most efficiently in slow-release pellets or capsules per os, though the method can be laborious. Mineral deficiencies in pasture or crops caused by excessive soil acidity or alkalinity are best controlled and prevented by correcting soil pH, by applying or withholding lime, dolomite or sulfate fertilizers. Examples include manganese deficiency caused by excessive liming (Wilson, 1952, 1965) and selenium deficiency in pasture grown in very acid soils which responds to fertilization with dolomite or lime. This method is the best for the reason that the correction of soil pH aids pasture/crop growth and therefore potential animal production.

The recommended methods are shown in Table 6.2. The rates are guides only and the responses should be measured chemically (soils, pastures) and by clinical pathology.

The common practice of feeding dairy cows on a combination of pasture (paddock or zero-grazed) and concentrates creates some difficulties in respect to treatment control and prevention. Ideally, the amount and composition (i.e. formulation) of the concentrate component of the total diet are altered daily to compensate for variations in the quantity and composition of the pasture. This is possible only when the manager is able to estimate accurately these variables and where the compounding of the concentrates is done on the farm. In practice, it is possible with some loss of efficiency, to compensate for variations in the pasture component of the ration by adjusting the quantity of the standard concentrate components of the ration offered. Increased quantities are given when the pastures (or crops) are immature and when they are maturing and becoming indigestible. At later stages of growth it is necessary to reduce the quantity of pasture offered to allow cows at the peak of lactation to eat sufficient DOM. It is also necessary to increase the concentration of any minerals in the concentrates which are deficient in the pasture.

When by-products are to be included in feedstuffs for breeding cows, care needs to be taken to ensure that they are free of contaminants such as heavy metals, insecticides and other chemical poisons, from toxic fungi and pathogenic bacteria and their toxins, and are non-toxic. They must also be nutritionally suitable for the class of cattle concerned.

Concentrates

Concentrates formulated and compounded by reputable feedstuff manufacturers should contain the ingredients and the correct analysis as stated on the label. Care needs to be taken in formulating and mixing concentrates on-farm, so that the product meets the requirements of the cows when fed in conjunction with the roughage component.

The Very High-producing Cow

Cows which have been bred for very high milk yields (and possibly those treated with bovine somatotrophin) present some difficulties. There seem

Table 6.2. Mineral fertilizers, supplements and therapeutics for treatment, control and prevention of bovine infertility.

Mineral	Pasture fertilizer	Feed supplement	Therapeutic	Remarks
Cobalt	300–600 g $CoSO_4$ ha^{-1} $year^{-1}$	0.07 mg kg^{-1} DM	1 × 30 g slow-release pellet, 30% Co, 70% Fe per os at 6-monthly intervals (1)	
Copper	Up to 5 kg $CuSO_4$ ha^{-1} $year^{-1}$	10 mg kg^{-1} DM	50 g Cu in CuO needles in capsule (2) at 6–12-monthly intervals per os (3)[a]	Response should be monitored
Iodine		0.8–1.0 mg kg^{-1} DM or salt licks containing 150–200 g potassium iodate per tonne		
Manganese	Correct soil pH	4 g $MnSO_4$ day^{-1} (4)		
Phosphorus	Up to 1000 kg super-phosphate ha^{-1} $year^{-1}$ (usually up to 200 kg)	Dicalcium phosphate (5) or steamed bone flour 50 g day^{-1}, or licks containing steamed bone flour in molasses or salt, or monosodium dihydrogen phosphate in drinking water 0.5–1.0 g l^{-1}		

Selenium	10 g Se ha^{-1} year^{-1} in encapsulated form, or correct soil pH	0.1 mg Se kg^{-1} DM as sodium selenate/ite	2 × 30 g slow-release pellets containing 10% Se, 90% Fe (6) four weeks before joining heifers then annually at drying off[a]	Response *must* be monitored

[a] Alternatively, and in special circumstances, copper and selenium can be administered parenterally (7), though twice or thrice annual injections of drugs cannot be regarded as good animal husbandry practice. Copper – as copper methionate, glycinate, calcium EDTA, or diethylamine copper oxyquinoline sulfonate, at the rate of up to 1 mg Cu kg^{-1} bodyweight s.c. at 4-6-monthly intervals. Selenium – as sodium selenate or selenite at the rate of up to 0.15 mg Se kg^{-1} bodyweight s.c. at 4-monthly intervals.

(1) Reid and McQueen (1985).
(2) Suttle (1981); Gallagher and Cottrill (1985); Deland *et al.* (1986).
(3) MacPherson (1985).
(4) Wilson (1966) (>4 g may be necessary where deficiency is severe).
(5) Arthur *et al.* (1989).
(6) Judson *et al.* (1980); McClure *et al.* (1986).
(7) Allen and Mallinson (1984).

to be four possible courses of action in the treatment, control and prevention of infertility in these cows.

1. To increase the energy content of the rations, e.g. by adding protected fats (Lucy et al., 1991; Sklan et al., 1991), although these have given erratic responses (Lucy et al., 1992b).
2. To administer gonadotrophin-releasing hormone (GnRH) where the pathogenesis involves pituitary gonadotrophic failure and progesterone where there is primary ovarian failure.
3. To limit peak milk yield by controlling liveweight and condition pre-partum.
4. To accept the fact that very high-producing cows will have longer calving-to-conception intervals and inter-calving intervals of more than 365 days.

Breeders of dairy cattle need to consider amongst their selection criteria the capacity of cows to ingest and absorb sufficient nutrients to supply the energy for high milk yields. Producers need to balance their wishes for high yields with high fertility, a decision which usually will be based on economic grounds.

Monitoring

Competent herd managers are usually able to achieve high levels of herd fertility by close attention to the intake and quality of feed, liveweight and condition of the cows and, in dairy herds, yield and composition of the milk.

Techniques are available for measuring these and for obtaining additional data for regular analysis. Such analyses (usually by computer) should indicate the existence of any incipient nutritional and metabolic causes of infertility, enabling remedial measures to be taken before major economic losses occur.

The level of monitoring practised is a matter for managerial decision after weighing the probability and economic consequences of nutritional and metabolic infertility.

Nutrition

The data required are: feed type; daily DM intake; chemical composition (DM, ME, readily available carbohydrate, fibre, crude protein (NPN, rumen-degradable protein, UDP); cobalt, copper, iodine, manganese, molybdenum, phosphorus, sulfur and selenium; β-carotene and α-tocopherol.

These data are easily obtained for hand-fed cattle. Accurate measures of intake and feed composition are impossible to obtain for grazing cattle because of sampling errors and the continual change in stage of pasture growth and composition. The following methods have been found to yield reasonable estimates.

1. Intake. Though accurate intake measurements of available herbage mass can be made using methods described by Frame (1981), simpler methods for monitoring feed intake which are applicable in the field exist.

Under strip-grazing conditions, where there is a large amount of pasture on offer and a considerable amount removed during grazing, the daily dry matter intake of grazing cattle can be determined with a satisfactory degree of accuracy by measuring the DM in the pasture before grazing, with a grassmeter such as the Massey Grassmeter (Holmes, 1974), the Ellinbank Rising Plate Grassmeter (McGowan and Earle, 1978; Earle and McGowan, 1979), or the electronic capacitance meters described by Campbell et al. (1962) and Vickery et al. (1980), and again after grazing.

2. Composition. At the time the available herbage mass is being measured, observations are made of the botanical composition of the pasture and stage of maturity, e.g. 10% white clover, 90% perennial ryegrass, both at the early flowering stage of growth. Periodically, samples of pasture are taken from the locations measured by the grassmeter, dried and analysed. It is possible with experience and occasional sampling, to estimate the probable amount and composition of pasture (Parker, 1973) before and after grazing and so deduce the approximate mean intake of each of the essential nutrients.

The intake of essential nutrients is then compared with the cows' requirements either on an individual or group basis.

Deficiencies in the intake of essential nutrients are likely to be followed by some change in blood metabolite concentrations, pool-size or rate of transfer, other blood parameters, loss of weight and condition and some degree of infertility in susceptible animals. As these changes can occur within days of change in feed intake and composition, it is important that they be detected early.

Clinical signs

The desirable liveweight and liveweight changes of cattle are:

Weaning to puberty: 0.5 kg day^{-1} (Jersey), 0.6 kg day^{-1} (Holstein-Friesian) (Reid et al., 1964; Schultz, 1969; Roy, 1980). See also Table 4.1.
Late pregnancy: approximately 0.5 kg day^{-1}. At term, the liveweight increase should have at least equalled the weight of the fetus, membranes and fluids, approximately 15% of total liveweight (Putnam and Henderson, 1946).
Parturition to mating: Nil weight loss, or if a weight loss has occurred, a weight gain over the last three weeks before mating.

The desirable condition score at calving is 2.5 to 3 on the scale 1–5, or 5 to 6 on the 1–8 or 1–9 scales, and at mating ⩾2.5 or ⩾5 respectively (Kilkenny, 1978, 1982; Rae et al., 1993).

No signs of malnutrition (achromotrichia, lameness, etc.) are permitted.

Clinical pathology

Milk and blood samples are taken from groups of cows ($n = 7$–10) at weekly intervals over the critical stages of reproduction, particularly over the mating period from three weeks before mating until three weeks after. The samples are analysed for the relevant nutrients, metabolites and hormones, the most important which appear to be:

1. **Blood.** Glucose, β-hydroxybutyrate, free fatty acids, albumin, P and Se, and progesterone (oestrus −8 to 15 days). The mean values and standard deviations for each component are determined and compared with the normal values (population mean ±2 SD). Values which are outside the normal range or which are falling at the time of mating, and progesterone 8–15 days before mating of <5 ng ml^{-1}, may be significant. The normal ranges are shown in Table 6.3.
2. **Milk.** Yield, fat and protein.

Heat Detection

Optimal nutrition is necessary to ensure the maximal expression of overt oestrus in herds where artificial insemination or hand service is practised (Suzuki et al., 1982).

Table 6.3. Concentrations of nutrients and metabolites in the blood of adult cattle of possible value in monitoring metabolic activity for reproductive purposes.

	Mean	Standard deviation	Reference
Packed cell volume (l l^{-1})	0.30	0.03	(1)
Haemoglobin (g dl^{-1})	12.1	1.0	(1)
Glucose (blood, mmol l^{-1})	2.5	0.25	(1)
Albumin (serum, g l^{-1})	34	2.5	(1)
Urea (serum, mmol l^{-1})	2.4	0.42	(1)
Cobalt (methylmalonic acid) (serum, μmol l^{-1})	>2.0		(4)
Copper (blood, μmol l^{-1})	14.13	2.67	(3)
(serum, μmol l^{-1})	12.56	2.67	(3)
Iodine (serum total, μmol l^{-1})	732.8	319.7	(2)
Phosphorus (serum inorganic, mmol l^{-1})	2.0	0.27	(1)

(1) Payne et al. (1973).
(2) Hewett (1974).
(3) Mylrea and Bayfield (1968).
(4) McGhie (1991).

The detection of oestrus has been described by Esslemont (1975) and McLeod and Williams (1991). Heat can be made more obvious by the use of heat-mount detectors such as coloured pressure pads and tail paste (Baker, 1965; Ball et al., 1983; Thibier et al., 1983).

Hormone Therapy

Techniques have been described for various forms of hormone therapy to treat infertile cows. These include pregnant mare's serum gonadotrophin (PMSG), progesterone-impregnated intravaginal device, human chorionic gonadotrophin (hCG), GnRH and progesterone (Danieli, 1967; Grunert and Diez, 1976; Lamming and Bulman, 1976; Bulman and Lamming, 1978; Bulman et al., 1978; Moller and Fielden, 1981; Macmillan et al., 1991; Mee et al., 1993; Morgan and Lean, 1993).

Reservations are expressed as to the justification of hormonal therapy for nutritional and metabolic infertility on rational, physiological, economic and ethical grounds.

Weaning

Weaning has a role in embryo transfer operations as it removes two of the important predisposing causes of metabolic infertility, milk secretion and suckling, and hastens the return of hypothalamo-pituitary–ovarian function (Reisen et al., 1968; Laster et al., 1973; Smith et al., 1979; Lusby and Wettemann, 1980; Tervit et al., 1982; Walters et al., 1982). For example, Laster et al. (1973) raised the proportion of two-year-old cows showing oestrous cycles from 63.0 to 92.0%, three-year-old cows from 70.0 to 96.7% and adult cows from 81.4 to 97.7% during a 42-day breeding period commencing, on average, 65 days after calving, by weaning calves at an average age of 55 days.

Conclusions

Nutritional and metabolic factors are just two of the many causes of reproductive failure in cattle and should not be considered in isolation when developing control and preventive programmes. Techniques for these should be incorporated into programmes which will identify, control and prevent all causes of herd infertility before reproduction is affected. Some indications as to how such programmes can be developed are given in this example for extensively managed, seasonally mated herds.

1. Two weeks before the commencement of joining the bulls with the cows, obtain details of mating policy and practice, and the details of the herd management and nutrition and examine:

(a) The bulls for fertility, nutritional state and infection with *Campylobacter fetus*, *Tritrichomonas fetus* and pestivirus, infectious bovine rhinotracheitis (IBR), etc.

(b) A sample of the cows, including representatives of all age groups for lactational and nutritional state (liveweight and condition), ovarian activity and disease. While it is impossible to determine the state of ovarian activity in single cows at one examination, a good indication of the ovarian activity of a group of cows can be found at a single examination. Follicles remain palpable for about 24 hours and corpora lutea palpable for 12–14 days in each cycle. This indicates that follicles should be palpable in 1/21 or 5%, corpora lutea in 13/21 or 62%, and neither in 7/21 or 33% of the cows.

(c) The pastures that are being grazed and those that will be grazed during the breeding season for quantity and quality (composition), taking into account distances between watering points.

Depending upon the knowledge of the local soils and pastures and the results of clinical examination of the cows, samples may be taken from soils, pastures and cows for chemical examination.

2. During the first three or four weeks of mating, the bulls (or a sample of the bulls running with a sample of the cows) can be raddled, e.g. using a chin-ball harness, and the proportion of cows mated during the first three or four weeks determined.

3. Eight weeks after the bulls have been removed, cows and bulls are then re-examined and sampled as at the visit two weeks before the start of joining. At this visit the pregnancy rate of a sample of each class of animal is determined by rectal examination.

4. A fourth visit can be made four weeks before the beginning of the calving season to examine a sample of cows again for pregnancy and the abortion rate determined, and the cows examined again for evidence of disease.

5. Finally, the perinatal mortality, calving and weaning rates are determined.

Data obtained in the course of the investigation enables the reproductive performance to be predicted and identifies the causes of any reproductive failure that might be detected in time to take the appropriate remedial action to control or prevent economic loss.

References

Adler, J.H. and Trainin, D. (1961) The apparent effect of alfalfa on the reproductive performance of dairy cattle. *Proceedings of the IVth International Congress on Animal Reproduction* 3, 451–456.

AFRC (1992) AFRC Technical Committee on Responses to Nutrients Report No. 9. Nutritive Requirements of Ruminant Animals: Protein. *Nutrition Abstracts and Reviews* (Series B) 62, 787–835.

Ahlswede, L. and Lotthammer, K.H. (1978) Untersuchungen über eine spezifische, Vitamin-A-unabhängige Wirkung des β-Carotins auf die Fertilität des Rindes. 5. Mitteilung: Organuntersuchungen (Ovarien, Corpora lutea, Leber, Fettgewebe, Uterussekret, Nebennieren) – Gewichts – und Gehaltbestimmungen. *Deutsche Teirärztliche Wochenschrift* 85, 7–12.

Alderman, G. (1963) Mineral nutrition and reproduction in cattle. *Veterinary Record* 75, 1015–1018.

Alderman, G. (1970) Nutrition as a possible cause of infertility in cattle. *Veterinary Record* 87, 35.

Allan, C.L., Hemmingway, R.G. and Parkins, J.J. (1993) Improved reproductive performance in cattle dosed with trace element/vitamin boluses. *Veterinary Record* 132, 463–464.

Allcroft, R., Scarnell, J. and Hignett, S.L. (1954) A preliminary report on hypothyroidism in cattle and its possible relationship with reproductive disorders. *Veterinary Record* 66, 367–371.

Allen, W.M. and Mallinson, C.B. (1984) Parenteral methods of supplementation with copper and selenium. *Veterinary Record* 114, 451–454.

Anke, M., Groppel, B., Krauter, U. and Muller, M. (1991) [Effect of sulphur, cadmium and molybdenum pollution in ruminants and in man.] *Umweltaspekte der Tierproduktion* 33, 421–426 (in German).

Anon. (1983) *Selenium in Nutrition.* National Academy Press, Washington, DC, pp. 97–98.

ARC (Agricultural Research Council, London) (1980) *The Nutrient Requirements*

of Ruminant Livestock. Commonwealth Agricultural Bureaux, Farnham Royal, UK.
ARC (Agricultural Research Council, London) (1984) *The Nutrient Requirements of Ruminant Livestock, Supplement No. 1*. CAB International, Farnham Royal, UK.
Armstrong, J., Henderson, A.G., Lang, D.R., Robinson, D.W. and Suijendorp, H. (1968) Preliminary observations on the productivity of female cattle in the Kimberley region of north-western Australia. *Australian Veterinary Journal* 44, 357–363.
Armstrong, J.D., Goodall, E.A., Gordon, F.J., Rice, D.A. and McCaughey, W.J. (1990) The effects of levels of concentrate offered and inclusion of maize gluten or fish meal in the concentrate on reproductive performance and blood parameters of dairy cows. *Animal Production* 50, 1–10.
Arthur, G.H., Noakes, D.E. and Pearson, H. (1989) *Veterinary Reproduction and Obstetrics*, 6th edn. Baillière Tindall, London, p. 320.
Ascarelli, I., Edelman, Z., Rosenberg, M. and Folman, Y. (1985) Effect of dietary carotene on fertility of high yielding dairy cows. *Animal Production* 40, 195–207.
Asdell, S.A. (1964) *Patterns of Mammalian Reproduction*, 2nd edn. Cornell University Press, Ithaca, New York.
Axelsen, A., Bennett, D., Larkham, P.A. and Coulton, L. (1972) Effect of calving time and stocking rate on production of beef cows. *Proceedings of the Australian Society of Animal Production* 9, 165–170.
Ayalon, N. (1978) A review of embryonic mortality in cattle. *Journal of Reproduction and Fertility* 54, 483–493.
Baile, C.A., McLaughlin, C.L., Chalupa, W.V., Synder, D.L., Pendlum, L.C. and Potter, E.L. (1982) Effects of monensin fed to replacement dairy heifers during the growing and gestation period upon growth, reproduction and subsequent lactation. *Journal of Dairy Science* 65, 1941–1944.
Bailey, R.W. (1962) Pasture carbohydrates and the ruminant. *Proceedings of the New Zealand Society of Animal Production* 22, 99–108.
Baker, A.A. (1965) Comparison of heat mount detectors and classical methods for detecting heat in beef cattle. *Australian Veterinary Journal* 41, 360–361.
Ball, P.J.H., Cowpe, J.E.D. and Harker, D.B. (1983) Evaluation of tail paste as an oestrus detection aid using serial progesterone analysis. *Veterinary Record* 112, 147–149.
Baly, D.L., Keen, C.L., Curry, D.L. and Hurley, L.S. (1985) Effects of manganese deficiency on carbohydrate metabolism. *Trace Elements in Man and Animals* 5, 254–258.
Barber, W.P. and Lonsdale, C.R. (1980) By-products from cereal, sugar beet and potato processing. In: *By-Products and Wastes in Animal Feeding. Occasional Publication No. 3*. British Society of Animal Production, Reading, pp. 61–69.
Barr, N.C.E. (1971) *Reproduction in Queensland Beef Cattle*. Queensland Department of Primary Industries, Brisbane.
Barraclough, C.A. and Wise, P.M. (1982) The role of catecholamines in the regulation of pituitary luteinizing hormone and follicle-stimulating hormone secretion. *Endocrine Reviews* 3, 91–119.

Bath, I.H. and Rook, J.A.F. (1963) The evaluation of cattle foods and diets in terms of the ruminal concentration of volatile fatty acids. 1. The effects of level of intake, frequency of feeding, the ratio of hay to concentrates in the diet, and of supplementary feeds. *Journal of Agricultural Science* 61, 341–348.

Bazer, F.W., Thatcher, W.W., Hansen, P.J., Mirando, M.A., Ott, T.L. and Plante, C. (1991) Physiological mechanisms of pregnancy recognition in ruminants. *Journal of Reproduction and Fertility*, Supplement 43, 39–47.

Bedrak, E., Warnick, A.C., Hentges, J.F. and Cunha, T.J. (1964) Effect of protein intake on gains, reproduction and blood constituents of beef heifers. *Florida Agricultural Experiment Station, Technical Bulletin* No. 678.

Beguin, D.P., Kincaid, R.L. and Hargis, A.M. (1985) Fetal death in copper-deficient rats. *Nutrition Reports International* 31, 991–998.

Bennet, R.C., Seath, D.M. and Olds, D. (1968) Relationship between nitrate content of forage and dairy herd fertility. *Journal of Dairy Science* 51, 629.

Bentley, O.G. and Phillips, P.H. (1951) The effect of low manganese rations upon dairy cattle. *Journal of Dairy Science* 34, 396–403.

Biggers, J.D. and Borland, R.M. (1976) Physiological aspects of growth and development of the preimplantation mammalian embryo. *Annual Reviews of Physiology* 38, 95–119.

Bines, J.A. and Davey, A.W.F. (1970) Voluntary intake, digestion, rate of passage, amount of material in the alimentary tract and behaviour in cows receiving complete diets containing straw and concentrates in different proportions. *British Journal of Nutrition* 24, 1013–1028.

Bing, R.J. (1965) Cardiac metabolism. *Physiological Reviews* 45, 171–213.

Blanchard, T., Ferguson, J., Love, L., Takeda, T., Henderson, B., Hasler, J. and Chalupa, W. (1990) Effect of dietary crude-protein type on fertilization and embryo quality in dairy cattle. *American Journal of Veterinary Research* 51, 905–908.

Blauwiekel, R., Kincaird, R.L. and Reeves, J.J. (1986) Effect of high crude protein on pituitary and ovarian function in Holstein heifers. *Journal of Dairy Science* 69, 439–446.

Blaxter, K.L., Wainman, F.W. and Wilson, R.S. (1961) The regulation of food intake by sheep. *Animal Production* 3, 51–61.

Block, E. (1991) Anion–cation balance and its effect on the performance of ruminants. In: Haresign, W. and Cole, D.J.A. (eds) *Recent Advances in Animal Nutrition*. Butterworth–Heinemann, Oxford, pp. 163–179.

Blood, D.C., Radostits, O.M. and Henderson, J.A. (1983) *Veterinary Medicine*, 6th edn. Baillière Tindall, London.

Blum, J.M., Gingins, M., Schnyder, W., Kunz, P., Thomson, E.F., Vitins, P., Blom, A., Burger, A. and Bickel, H. (1979) Energy intake in ruminants: effects on blood plasma levels of hormones and metabolites. *International Journal for Vitamin and Nutrition Research* 49, 121–122.

Bogin, E., Avidar, Y., Davidson, M., Gordin, S. and Israeli, B. (1982) Effect of nutrition on fertility and blood composition in the milk cow. *Journal of Dairy Research* 49, 13–23.

Bond, J., Wiltbank, J.N. and Cook, A.C. (1958) Cessation of estrus and ovarian activity in a group of beef heifers on extremely low levels of energy and protein. *Journal of Animal Science* 17, 1211.

Bonomi, A., Quarantelli, A., Sabbioni, A., Superchi, P., Bosticco, A., Chiesa, F. and Gaiani, R. (1989) Contributo allo studio dell'influenza esercitata dalla methionina protetta, impiegata nel ruolo di supplemento alimentare, sull'efficienza produttiva e riproductiva nelle bovine da latte. *Revista della Societa Italiana di Scienza dell'Alimentazione* 18, 273–288.

Boyd, H., Ritchie, N.S., Cooke, B.C. and Roche, J.F. (1984) Beta-carotene status of in-wintered cattle. *Irish Veterinary Journal* 38, 95–100.

Brinster, R.L. (1967) Carbon dioxide production from lactate and pyruvate by the preimplantation mouse embryo. *Experimental Cell Research* 47, 634–637.

Brinster, R.L. (1970) Culture of two-cell rabbit embryos to morulae. *Journal of Reproduction and Fertility* 21, 17–22.

Broster, W.H. (1973) Liveweight change and fertility in the lactating dairy cow: a review. *Veterinary Record* 93, 417–420.

Bryant, A.M. (1965) The effect of nitrate on the *in vitro* fermentation of glucose by rumen liquor. *New Zealand Journal of Agricultural Research* 8, 118–125.

Bryant, A.M. and Trigg, T.E. (1982) The nutrition of the grazing dairy cow in early lactation. In: Macmillan, K.L. and Taufa, V.K. (eds) *Proceedings of the Conference on Dairy Production from Pasture.* New Zealand and Australian Societies of Animal Production, Ruakura Animal Research Station, Hamilton, New Zealand, pp. 185–207.

Bryant, A.M. and Ulyatt, M.J. (1965) Effects of nitrogenous fertilizer on the chemical composition of short-rotation ryegrass and its subsequent digestion by sheep. *New Zealand Journal of Agricultural Research* 8, 109–117.

BSAP (1980) *By-Products and Wastes in Animal Feeding. Occasional Publication No. 3.* British Society of Animal Production, Reading.

Buchanan-Smith, J.G., Bannister, W., Durham, R.M. and Curl, S.E. (1964) Effect of all-concentrate fed *ad libitum* versus roughage ration on occurrence of estrus in beef heifers. *Journal of Animal Science* 23, 902.

Bulman, D.C. and Lamming, G.E. (1978) Milk progesterone levels in relation to conception, repeat breeding and factors influencing acyclicity in dairy cows. *Journal of Reproduction and Fertility* 54, 447–458.

Bulman, D.C., McKibbin, P.E., Appleyard, W.T. and Lamming, G.E. (1978) Effect of a progesterone-releasing intravaginal device on the milk progesterone levels, vaginal flora, milk yield and fertility of cyclic and non-cyclic dairy cows. *Journal of Reproduction and Fertility* 53, 289–296.

Butler, G.W. and Johnston, J.M. (1957) Factors influencing the iodine content of pasture herbage. *Nature (London)* 179, 216.

Butler, G.W. and Peterson, P.J. (1961) The effects of different pasture species and varieties on sheep production. IV. Some aspects of the chemical composition of pasture herbage In: *Sheep Farming Annual.* Massey College, Palmerston North, New Zealand, pp. 187–194.

Butler, W.R. and Smith, R.D. (1989) Interrelationships between energy balance and postpartum reproductive function in dairy cattle. *Journal of Dairy Science* 72, 767–783.

Butler, W.R., Everett, R.W. and Coppock, C.E. (1981) The relationships between energy balance, milk production and ovulation in postpartum Holstein cows. *Journal of Animal Science* 53, 742-748.

Campbell, A.G. (1982) Dairyfarm production and profitability from maize/pasture rotation systems. In: Macmillan, K.L. and Taufa, V.K. (eds) *Proceedings of the Conference on Dairy Production from Pasture.* New Zealand and Australian Societies of Animal Production, Ruakura Animal Research Station, Hamilton, New Zealand, pp. 240-241.

Campbell, A.G., Phillips, D.S.M. and O'Reilly, E.D. (1962) An electronic instrument for pasture yield estimation. *Journal of the British Grassland Society* 17, 89-100.

Campbell, A.G., Clayton, D.G. and Bell, B.A. (1977) Milkfat production from No. 2 Dairy, Ruakura. How it is attained: what it is worth. *New Zealand Journal of Agricultural Science* 11, 73-86.

Campling, R.C. (1964) Factors affecting the voluntary intake of grass. *Proceedings of the Nutritional Society* 23, 80-88.

Campling, R.C., Freer, M. and Balch, C.C. (1962) Factors affecting the voluntary intake of food by cows. 3. The effect of urea on the voluntary intake of oat straw. *British Journal of Nutrition* 16, 115-124.

Canfield, R.W. and Butler, W.R. (1991) Energy balance, first ovulation and the effects of naloxone on LH secretion in early postpartum dairy cows. *Journal of Animal Science* 69, 740-746.

Carroll, D.J., Barton, B.A., Anderson, G.W. and Smith, R.D. (1988) Influence of protein intake and feeding strategy on reproductive performance of dairy cows. *Journal of Dairy Science* 71, 3470-3481.

Cohen, R.D.H. (1975) Phosphorus for grazing beef cattle. *Australian Meat Research Committee Review* 23, 1-16.

Connor, H.C., Houghton, P.L., Lemenager, R.P., Malven, P.V., Parfet, J.R. and Moss, G.E. (1990) Effect of dietary energy, body condition and calf removal on pituitary gonadotropins, gonadotropin-releasing hormone (GnRH) and hypothalamic opioids in beef cows. *Domestic Animal Endocrinology* 7, 403-411.

Cooke, B.C. and Pugh, M.L. (1980) Slaughter waste in animal feeding. In: *By-Products and Wastes in Animal Feeding. Occasional Publication No. 3.* British Society of Animal Production, Reading, pp. 79-83.

Cooper, K.J., Fawcett, C.P. and McCann, S.M. (1974) Variations in pituitary responsiveness to a luteinizing hormone/follicle stimulating hormone releasing factor (LH-RF/FSH-RF) preparation during the rat estrous cycle. *Endocrinology* 95, 1293-1299.

Corbett, J.L. (1980) Grazing ruminants: evaluation of their feeds and needs. *Proceedings of the New Zealand Society of Animal Production* 40, 136-144.

Corbett, J.L., Langlands, J.P. and Reid, G.W. (1963) Effects of season of growth and digestibility of herbage on intake by grazing dairy cows. *Animal Production* 5, 119-129.

Crichton, J.H., Aitkin, J.N. and Boyne, A.W. (1959) Tho effect of plane of nutrition during rearing on growth, production, reproduction and health of dairy cattle. 1. Growth to 24 months. *Animal Production* 1, 145-162.

Cross, B.A and Dyer, R.G. (1972) Ovarian modulation of unit activity in the anterior hypothalamus of the cyclic rat. *Journal of Physiology (London)* 222, 25 pp.

Danieli, Y. (1967) Treatment of "repeat breeder" cows with progesterone. *Refuah Veterinarith* 24, 38–36.

Danilin, G.V. (1966) [Effect of I on fertility of barren cows and thyroid function in a region of I deficiency. *Zivotnovodstvo* 9, 87–88 (in Russian).

Davidson, M., Francos, G. and Meir, E. (1978) An analysis of several nutritional parameters in kibbutz dairy herds with low fertility. *Refuah Veterinarith* 35, 170–167.

Davies, H.L. (1962) Intake studies in sheep involving high fluid intake. *Proceedings of the Australian Society of Animal Production* 4, 167–171.

Davies, S., Baird, F. and Aizlewood, S. (1992) Use of free-access minerals. *Veterinary Record* 131, 131–132.

Deland, M.P.B., Lewis, D., Cunningham, P.R. and Dewey, D.W. (1986) Use of orally administered oxidised copper wire particles for copper therapy in cattle. *Australian Veterinary Journal* 63, 1–3.

Dindorkar, C.V., Muzaffar, S.J. and Kaikini, A.S. (1982) Note on the effect of feeding extra energy diets to anoestrous cattle. *Indian Journal of Animal Science* 52, 1228–1229.

Diplock, A.T. (1983) In: Porter, R. and Whelan, J. (eds) *Biology of Vitamin E*. Pitman, London, pp. 45–55.

Diskin, M.G. and Sreenan, J.R. (1980) Fertilization and embryonic mortality rates in beef heifers after artificial insemination. *Journal of Reproduction and Fertility* 59, 463–468.

Dodson, M.E. and Judson, G.J. (1973) Low blood selenium concentrations in cattle. *Australian Veterinary Journal* 49, 320.

Donaldson, L.E. (1962) Some observations on the fertility of beef cattle in North Queensland. *Australian Veterinary Journal* 38, 447–454.

Donaldson, L.E. (1966) Simple home-mixed feed supplements for range cattle. *Australian Veterinary Jornal* 42, 245–246.

Donaldson, L.E. and Hansel, W. (1964) The nature of the luteotrophic hormone in the bovine. In: *Proceedings of the 5th International Congress on Animal Reproduction and Artificial Insemination* No. 3, pp. 347–350.

Donaldson, L.E., Bassett, J.M. and Thorburn, G.D. (1970) Peripheral plasma progesterone concentration of cows during puberty, oestrous cycles, pregnancy and lactation, and the effects of undernutrition or exogenous oxytocin on progesterone concentration. *Journal of Endocrinology* 48, 599–614.

Downie, J.G. and Gelman, A.L. (1976) The relationship between changes in bodyweight plasma glucose and fertility in beef cows. *Veterinary Record* 99, 210–212.

Downing, J.A. and Scaramuzzi, R.J. (1991) Nutrient effects on ovulation rate, ovarian function and secretion of gonadotrophic and metabolic hormones in sheep. *Journal of Reproduction and Fertility*, Supplement 43, 209–227.

Drouva, S.V., Laplante, E., Gautron, J.-P. and Kordon, C. (1984) Effects of 17β-estradiol on LH-RH release from rat mediobasal hypothalamic slices. *Neuroendocrinology* 38, 152–157.

Ducker, M.J., Yarrow, N.H., Bloomfield, G.A. and Edwards-Webb, J.D. (1984)

The effect of β-carotene on the fertility of dairy heifers receiving maize silage. *Animal Production* 39, 9–16.

Ducker, M.J., Haggett, R.A., Fisher, W.J., Morant, S.V. and Bloomfield, G.A. (1985a) Nutrition and reproductive performance of dairy cattle. 1. The effect of level of feeding in late pregnancy and around the time of insemination on the reproductive performance of first lactation dairy heifers. *Animal Production* 41, 1–12.

Ducker, M.J., Morant, S.V., Fisher, W.J. and Haggett, R.A. (1985b) Nutrition and reproductive performance of dairy cattle. 2. Prediction of reproductive performance in first lactation dairy heifers subjected to controlled nutritional regimes. *Animal Production* 41, 13–22.

Dunn, T.G., Ingalls, J.E., Zimmerman, D.R. and Wiltbank, J.N. (1969) Reproductive performance of 2-year-old Hereford and Angus heifers as influenced by pre- and post-calving energy intake. *Journal of Animal Science* 29, 719–726.

Dyer, I.A. and Rojas, M.A. (1965) Manganese requirements and functions in cattle. *Journal of the American Veterinary Medical Association* 147, 1393–1396.

Dyer, R.G. (1985) Neural signals for oxytocin and LH release. *Oxford Reviews of Reproductive Biology* 7, 223–259.

Dyer, R.G. and McClure, T.J. (1981) Fasting inhibits oestrogen-stimulated secretion of luteinizing hormone in female rats. *Journal of Physiology (London)* 319, 91–92.

Dyer, R.G., Mansfield, S., Corbet, H. and Dean, A.D.P. (1985) Fasting impairs LH secretion in female rats by activating an inhibitory opioid pathway. *Journal of Endocrinology* 105, 91–97.

Earle, D.F. (1976) A guide to scoring dairy cow condition. *Journal of Agriculture (Victoria)* 74, 228–231.

Earle, D.F. and McGowan, A.A. (1979) Evaluation and calibration of an automatic rising plate meter for estimating dry matter yield of pasture. *Australian Journal of Experimental Agriculture and Animal Husbandry* 19, 337–343.

Eckles, C.H., Palmer, L.S., Gullickson, T.W., Fitch, C.P., Boyd, W.L., Bishop, L. and Nelson, J.W. (1935) Effects of uncomplicated phosphorus deficiency on estrus cycle, reproduction and composition of tissues of mature dairy cows. *Cornell Veterinarian* 25, 22–43.

Eger, S., Drori, D., Kadoori, I., Miller, N. and Schindler, H. (1985) Effects of selenium and vitamin E on incidence of retained placenta. *Journal of Dairy Science* 68, 2119–2122.

Erickson, G.F., Hofeditz, C. and Hsueh, A.J.W. (1983) GnRH stimulates meiotic maturation in preantral follicles of hypophysectomized rats. In: Greenwald, G.S. and Terranova, P.F. (eds) *Factors Regulating Ovarian Function*. Raven Press, New York, pp. 257–261.

Esslemont, R.J. (1975) The detection of oestrus in dairy cows. *Veterinary Annual* 15, 50–53.

Fagbohun, C.F. and Downs, S.M. (1992) Requirement for glucose in ligand-stimulated meiotic maturation of cumulus cell-enclosed mouse oocytes. *Journal of Reproduction and Fertility* 96, 681–697.

Fell, B.F., Dinsdale, D. and Mills, C.F. (1975) Changes in enterocyte mitochondria associated with deficiency of copper in cattle. *Research in Veterinary Science* 18, 274–281.

Ferguson, J.D. and Chalupa, W. (1989) Impact of protein nutrition on reproduction in dairy cows. *Journal of Dairy Science* 72, 746–766.

Ferguson, J.D., Blanchard, T.L. and Chalupa, W. (1986) The effects of protein level on reproduction in the dairy cow. In: *Proceedings of the Annual Meeting of the Society for Theriogenology*, 17–19 September 1986, Rochester, NY, pp. 164–185.

Fernando, G.W.E. and Carter, O.G. (1970) The effect of level of nitrogen fertilizer applied to forage oats on the grazing behaviour of dairy cattle. In: *Proceedings of the 11th International Grasslands Conference*, Surfers' Paradise, Queensland, pp. 853–856.

Filmer, J.F. (1933) Enzootic marasmus of cattle and sheep. Preliminary report having special reference to iron and liver therapy. *Australian Veterinary Journal* 9, 163–179.

Filmer, J.F. and Underwood, E.J. (1934) Enzootic marasmus. Treatment with limonite fractions. *Australian Veterinary Journal* 10, 83–87.

Filmer, J.F. and Underwood, E.J. (1937) Enzootic marasmus. Further data concerning the potency of cobalt as a curative and prophylactic agent. *Australian Veterinary Journal* 13, 57–64.

Folman, Y., Rosenberg, M., Herz, Z. and Davidson, M. (1973) The relationship between plasma progesterone concentration and conception in *post-partum* dairy cows maintained at two levels of nutrition. *Journal of Reproduction and Fertility* 34, 267–278.

Folman, Y., Ascarelli, I., Herz, Z., Rosenberg, M., Davidson, M. and Halevi, A. (1979) Fertility of dairy heifers given a commercial diet free of β-carotene. *British Journal of Nutrition* 41, 353–359.

Foot, A.S., Line, C. and Rowland, S.J. (1963) The effect of pre-partum feeding of heifers on milk composition. *Journal of Dairy Research* 30, 403–409.

Forbes, J.M. (1986) *The Voluntary Food Intake of Farm Animals.* Butterworths, London.

Foster, J.E., Biswell, H.H. and Hostetler, E.H. (1945) Comparison of different amounts of protein supplement for wintering beef cows on forest range in the Southeastern Coastal Plain. *Journal of Animal Science* 4, 387–394.

Frame, J. (1981) Herbage mass. In: *Sward Measurements Handbook.* British Grasslands Society, Maidenhead, Berkshire, pp. 39–69.

Francis, G.H. (1980) Use of vegetable and arable by-products. In: *By-Products and Wastes in Animal Feeding. Occasional Publication No. 3.* British Society of Animal Production, Reading, pp. 33–43.

Francos, G., Davidson, M. and Mayer, E. (1977) The influence of some nutritional factors on the incidence of the repeat breeder syndrome in high producing dairy herds. *Theriogenology* 7, 105–111.

Franklin, M.C. (1959) Factors affecting the performance of beef cattle on unimproved pastures in Queensland. *Australian Veterinary Journal* 35, 135–140.

Franzos, G. (1968) The relationship between the milk fat percentage and fertility in dairy herds. *Refuah Veterinarith* 25, 32–28.

Gallagher, J. and Cottrill, B.R. (1985) Methods of copper supplementation to cattle. *Veterinary Record* 117, 468.

Gartner, R.J.W., McLean, R.W., Little, D.A. and Winks, L. (1980) Mineral deficiencies limiting production of ruminants grazing tropical pastures in Australia. *Tropical Grasslands* 14, 266–272.

Gartner, R.J.W., Murphy, G.M. and Hoey, W.A. (1982) Effects of induced, subclinical phosphorus deficiency on feed intake and growth of beef heifers. *Journal of Agricultural Science (Cambridge)* 98, 23–29.

Gauthier, D., Terqui, M. and Mauleon, P. (1983) Influence of nutrition in pre-partum plasma levels of progesterone and total oestrogens and post-partum plasma levels of luteinizing hormone and follicle stimulating hormone in suckling cows. *Animal Production* 37, 89–96.

Girou, R. and Brochart, M. (1970a) Niveau énergétique, protéique et fécondité des vaches laitieres. Influence d'une supplémentation alimentaire post-oestrale. *Annales de Zootechnie* 19, 67–73.

Girou, R. and Brochart, M. (1970b) Effets d'une supplémentation alimentaire de brève durée sur le déclenchement des chaleurs chez des vaches en anoestrus post-partum. *Annales de Zootechnie* 19, 75–77.

Gleed, P.T., Allen, W.M., Mallinson, C.B., Rowlands, G.J., Sansom, B. and Vagg, M.J. (1983) Effects of selenium and copper supplementation on the growth of beef steers. *Veterinary Record* 113, 388–392.

Gombe, S. and Hansel, W. (1973) Plasma luteinizing hormone (LH) and progesterone levels in heifers on restricted energy intakes. *Journal of Animal Science* 37, 728–733.

Gordon, J.H. and Reichlin, S. (1974) Changes in pituitary responsiveness to luteinizing hormone-releasing factor during the rat estrous cycle. *Endocrinology* 94, 974–978.

Gordon, K., Renfree, M.D., Short, R.V. and Clarke, I.J. (1987) Hypothalamo-pituitary portal blood concentrations of β-endorphin during suckling in the ewe. *Journal of Reproduction and Fertility* 79, 397–408.

Goto, K., Kajisa, O., Ezoe, K., Nakanishi, Y., Ogaw, A.K., Tasaki, M., Ohta, H., Inohae, S., Tateyama, S. and Kawabata, T. (1989) Relationship between plasma beta-carotene concentration and embryo quality in super ovulated Japanese Black cattle. *Memoirs of the Faculty of Agriculture, Kagoshima University* 25, 113–117.

Grainger, C. and Wilhelms, G. (1979) Effect of duration and pattern of underfeeding in early lactation on milk production and reproduction of dairy cows. *Australian Journal of Experimental Agriculture and Animal Husbandry* 19, 395–401.

Graves-Hoagland, R.L., Hoagland, T.A. and Woody, C.O. (1988) Effects of beta-carotene and vitamin A on progesterone production by bovine luteal cells. *Journal of Dairy Science* 71, 1058–1062.

Greathead, K. (1983) The effects of calving date, liveweight and condition score on conception in mature lactating beef cows under grazing conditions. In: *Proceedings of Seminar on Reproduction in Farm Animals*. Australian Society of Animal Production, Department of Agriculture, Western Australia, pp. 71–75.

Greenhalgh, J.F.D. (1980) Use of straw and cellulose wastes and methods of improving their value. In: *By-Products and Wastes in Animal Feeding. Occasional Publication No. 3.* British Society of Animal Production, Reading, pp. 25–31.

Greenhalgh, J.F.D. and Runcie, K.V. (1962) The herbage intake and milk production of strip- and zero-grazed dairy cows. *Journal of Agricultural Science* 59, 95–102.

Griel, L.C., Patton, R.A., McCarthy, R.D. and Chandler, P.T. (1968) Milk production response to feeding methionine hydroxy analog to lactating dairy cows. *Journal of Dairy Science* 51, 1866–1868.

Grunert, E. and Diez, G. (1976) Zur Erhohung der Konzeptionsrate beim Rind durch HCG und Gn-RH. *Zuchthygiene* 11, 90.

Hacker, J.B. and Minson, D.J. (1981) The digestibility of plant parts. *Herbage Abstracts* 51, 459–482.

Halász, B. and Pupp, L. (1965) Hormone secretion of the anterior pituitary gland after physical interruption of all nervous pathways to the hypophysiotrophic area. *Endocrinology* 77, 553–562.

Haresign, W. (ed.) (1979) Body condition, milk yield and reproduction in cattle. In: *Recent Advances in Animal Nutrition.* Butterworths, London, pp. 107–122.

Harrison, J.H. and Conrad, H.R. (1982) Effect of feeding organic selenium (linseed meal) during the prepartum period for prevention of retained placenta, metritis and cystic ovaries. *Journal of Dairy Science* 65, Supplement P239, 182.

Harrison, J.H., Hancock, D.D. and Conrad, H.R. (1984) Vitamin E and selenium for reproduction of the dairy cow. *Journal of Dairy Science* 67, 123–132.

Harrison, R.O., Young, J.W., Freeman, A.E. and Ford, S.P. (1989) Effects of lactational level on reactivation of ovarian function, and interval from parturition to first visual oestrus and conception in high-producing Holstein cows. *Animal Production* 49, 23–28.

Hart, B. and Mitchell, G.L. (1965) Effect of phosphate supplementation on the fertility of an open range beef cattle herd on the Barkly Tableland. *Australian Veterinary Journal* 41, 305–309.

Hart, I.C., Bines, J.A., Morant, S.V. and Ridley, J.L. (1978) Endocrine control of energy metabolism in the cow: comparison of the levels of hormones (prolactin, growth hormone, insulin and thyroxine) and metabolites in the plasma of high- and low-yielding cattle at various stages of lactation. *Journal of Endocrinology* 77, 333–345.

Hartley, W.J. (1963) Selenium and ewe fertility. *Proceedings of the New Zealand Society of Animal Production* 23, 20–27.

Hecht, D., Wells, M.E., Bush, L.J. and Adams, G.D. (1977) Effects of dietary phosphorus levels on reproductive efficiency in dairy heifers. *Agricultural Research Station Reports, Oklahoma State University Miscellaneous Publication* 101, 126–129.

Heinonen, K., Sttala, E. and Alanko, M. (1988) Effect of postpartum live weight loss in reproductive functions in dairy cows. *Acta Veterinaria Scandinavica* 29, 249–254.

Helmer, S.D., Gross, T.S., Newton, G.R., Hansen, P.J and Thatcher W.W. (1989a) Bovine trophoblast protein-1 complex alters endometrial protein and prostaglandin secretion and induces an intracellular inhibitor of prostaglandin synthesis *in vitro. Journal of Reproduction and Fertility* 87, 421–430.

Helmer, S.D., Hansen, P.J., Thatcher, W.W., Johnson, J.W. and Bazer, T.W. (1989b) Intrauterine infusion of highly enriched bovine trophoblast protein-1 complex exerts an antiluteotrophic effect to extend corpus luteum lifespan in cyclic cattle. *Journal of Reproduction and Fertility* 87, 89–101.

Hewett, C. (1974) On the causes and effects of variations in the blood profile of Swedish dairy cattle. *Acta Veterinaria Scandinavica*, Supplement 50.

Hidiroglou, M. and Jenkins, K.J. (1968) Factors affecting the development of nutritional muscular dystrophy in northern Ontario. *Canadian Journal of Animal Science* 48, 7–14.

Hight, G.K. (1966) The effects of undernutrition in late pregnancy on beef cattle production. *New Zealand Journal of Agricultural Research* 9, 479–490.

Hight, G.K. (1968) Plane of nutrition effects in late pregnancy and during lactation on beef cows and their calves to weaning. *New Zealand Journal of Agricultural Research* 11, 71–84.

Hightshoe, R.B., Cochran, R.C., Corah, L.R., Kiracofe, G.H., Harmon, D.L. and Perry, R.C. (1991) Effects of calcium soaps of fatty acids on postpartum reproductive function in beef cows. *Journal of Animal Science* 69, 4097–4103.

Hignett, S.L. (1956) The influence of calcium, phosphorus, manganese and vitamin D on heifer fertility. In: *Proceedings of the 3rd International Congress on Animal Reproduction*, Plenary Sessions, p. 116.

Hignett, S.L. (1959) Some nutritional and other inter-acting factors which may influence the fertility of cattle. *Veterinary Record* 71, 247–256.

Hill, J.R., Lamond, D.R., Henricks, D.M., Dickey, J.F. and Niswender, G.D. (1970) The effects of undernutrition on ovarian function and fertility in beef heifers. *Biology of Reproduction* 2, 78–84.

Hino, T., Andoh, N. and Ohgi, H. (1993) Effects of β-carotene and α-tocopherol on rumen bacteria in the utilization of long chain fatty acids and cellulose. *Journal of Dairy Science* 76, 600–605.

Hixon, D.L., Fahey, G.C., Kesler, D.J. and Neumann, A.L. (1982) Effects of creep feeding and monensin on reproductive performance and lactation of beef heifers. *Journal of Animal Science* 55, 467–474.

Hodgson, J. (1982) Influence of sward characteristics on diet selection and herbage intake by the grazing animal. In: Hacker J.B. (ed.) *Nutritional Limits to Animal Production from Pastures.* Commonwealth Agricultural Bureaux, Farnham Royal, UK, pp. 153–166.

Hogan, J.P. and Weston, R.H. (1969) The digestion of pasture plants by sheep. III. The digestion of forage oats varying in maturity and in the content of protein and soluble carbohydrate. *Australian Journal of Agricultural Research* 20, 347–363.

Holmes, C.W. (1974) The Massey grassmeter. In: Rolston, S.J., Drummond, D.C., Williams, B. and Flux, D.S. (eds) *Dairy Farming Annual.* Massey University, Palmerston North, New Zealand, pp. 26–30.

Holmes C.W. and Macmillan, K.L. (1982) Nutritional management of the dairy herd grazing on pasture. In: Macmillan, K.L. and Taufa, V.K. (eds) *Proceedings of the Conference on Dairy Production from Pasture.* New Zealand

and Australian Societies of Animal Production, Ruakura Animal Research Station, Hamilton, New Zealand, pp. 244-274.

Holroyd, R.G., O'Rourke, P.K., Clarke, M.R. and Loxton, I.D. (1983) Influence of pasture type and supplement on fertility and liveweight of cows, and progeny growth rate in the dry tropics of northern Queensland. *Australian Journal of Experimental Agriculture and Animal Husbandry* 23, 4-13.

Howell, J.McC. and Davison, A.N. (1959) The copper content and cytochrome oxidase activity of tissues from normal and swayback lambs. *Biochemical Journal* 72, 365-368.

Howell, J.McC. and Hall, G.A. (1969) Histological observations on foetal resorption in copper-deficient rats. *British Journal of Nutrition* 23, 47-50.

Hulme, D.J, Kellaway, R.C., Booth, P.J. and Bennett, L. (1986) The CAMDAIRY model for formulating and analysing dairy cow rations. *Agricultural Systems* 22, 81-108.

Humblot, P., Camous, S., Martal, J., Charlery, J., Jeanguyot, N., Thibier, M. and Sasser, R.G. (1988) Pregnancy-specific protein B, progesterone concentrations and embryonic mortality during early pregnancy in dairy cows. *Journal of Reproduction and Fertility* 83, 215-223.

Humphrey, W.D., Kaltenbach, G.C., Dunn, T.G., Koritnik, D.R. and Niswender, G.D. (1983) Characterization of hormonal patterns in the beef cow during postpartum anestrus. *Journal of Animal Science* 56, 445-453.

Hunter, A.P. (1976) Nutritional and other factors affecting the fertility of dairy cattle in the Illawarra District on the south coast of New South Wales. MVSc Thesis, University of Sydney.

Hutton, J.B. (1961) Studies of the nutritive value of New Zealand dairy pastures. 1. Seasonal changes in some chemical components of pastures. *New Zealand Journal of Agricultural Research* 4, 583-590.

Hutton, J.B. (1962) The maintenance requirements of New Zealand dairy cattle. *Proceedings of the New Zealand Society of Animal Production* 22, 12-34.

Hutton, J.B. (1963) The effect of lactation on intake in the dairy cow. *Proceedings of the New Zealand Society of Animal Production* 23, 39-52.

Hutton, J.B. (1974) Recent researches on feeding for milk production. In: *Proceedings of the 19th International Dairy Congress*, pp. 280-294.

Hutton, J.B. and Parker, O.F. (1973) The significance of differences in levels of feeding, before and after calving, on milk yield under intensive grazing. *New Zealand Journal of Agricultural Research* 16, 95-104.

Hutton, J.B., Jury, K.E. and Davies, E.B. (1967) Studies of the nutritive value of New Zealand dairy pastures. V. The intake and utilization of potassium, sodium, calcium, phosphorus and nitrogen in pasture herbage by lactating dairy cattle. *New Zealand Journal of Agricultural Research* 10, 367-388.

Ishak, M.A., Larson, L.L., Owen, F.G., Lowry, S.R. and Erickson, E.D. (1983) Effects of selenium, vitamins and ration fiber on placental retention and performance of dairy cattle. *Journal of Dairy Science* 66, 99-106.

Jaskowski, J.M. and Rogoziewicz, M. (1990) [Influence of selenium administered post partum on ovarian disorders, conception rates and fertility in cows.] *Medcyna Weterynaryna* 46, 50-53 (in Polish).

Johns, A.T. (1962) Some aspects of rumen metabolism influencing intake and

production in sheep. *Proceedings of the New Zealand Society of Animal Production* 22, 88-98.
Jones, D.J.C. (1959) Studies of the chemical composition of kales and rapes. III. The minor elements. *Journal of Animal Science* 53, 151-155.
Jones, G.P., Garnsworthy, P.C. and Findlater, R.C.F. (1988) The effects of nutrition and body conditon at calving on the reproductive performance of dairy cows. In: *11th International Congress on Animal Reproduction and Artificial Insemination*, Dublin 2, Paper No. 35.
Jordan, E.R. and Swanson, L.V. (1979a) Serum progesterone and luteinizing hormone in dairy cattle fed varying levels of crude protein. *Journal of Animal Science* 48, 1154-1158.
Jordan, E.R. and Swanson, L.V. (1979b) Effect of crude protein on reproductive efficiency, serum total protein, and albumin in the high-producing dairy cow. *Journal of Dairy Science* 62, 58-63.
Judson, G.J., Mattschoss, K.H. and Clare, R.J. (1980) Selenium pellets for cattle. *Australian Veterinary Journal* 56, 304.
Kazmer, G.W., Barnes, M.A. and Canfield, R.W. (1985) Reproductive and metabolic hormones during estrus after fasting in Holstein heifers. *Theriogenology* 24, 619-629.
Kilkenny, J.B. (1978) Reproductive performance of beef cows. *World Review of Animal Production* 14(3), 65-74.
Kilkenny, J.B. (1982) Target condition scores for beef cows. *Animal Production* 34, 392.
King, G.J., Atkinson, B.A. and Robertson, H.A. (1979) Development of the bovine placentome during the second month of gestation. *Journal of Reproduction and Fertility* 55, 173-180.
King, J.O.L. (1968) The relationship between the conception rate and changes in bodyweight, yield and SNF content of milk in dairy cows. *Veterinary Record* 83, 492-494.
Klopfenstein, T. (1983) By-products for ruminants. In: *Proceedings of the 1983 Georgia Nutrition Conference for the Feed Industry*. University of Georgia, USA.
Kroker, G.A. and Cummins, L.J. (1979) The effect of nutritional restriction on Hereford heifers in late pregnancy. *Australian Veterinary Journal* 55, 467-474.
Krolak, M. (1968) Effect of manganese, added to diet, on cattle fertility and manganese content in hairs. *Polskie Archiwum Weterynaryjne* 11, 293-304.
Kronfeld, D.S. and Raggi, F. (1964) Glucose kinetics in normal, fasting, and insulin-treated cows. *American Journal of Physiology* 206, 109-112.
Kuswandi and Teleni, E. (1987) Glucose metabolism in growing lambs fed formaldehyde-treated casein supplements. *Proceedings of the Nutrition Society of Australia* 12, 177.
Kuzainov, K. and Kozhabergenov, O. (1988) [Lysine in diets for high producing cows.] *Vestnik Sel'Shokhozyastvennoi Nauki Kazakhstana* 10, 49-51 (in Russian).
Lamming, G.E. and Bulman, D.C. (1976) The use of milk progesterone radioimmunoassay in the diagnosis and treatment of subfertility in dairy cows. *British Veterinary Journal* 132, 507-517.

Lamming, G.E., Wathes, D.C. and Peters, A.R. (1981) Endocrine patterns of the post-partum cow. *Journal of Reproduction and Fertility*, Supplement 30, 155-170.

Larson, L.L., Mabruck, H.S. and Lowry, S.R. (1980) Relationship between early postpartum blood composition and reproductive performance in dairy cattle. *Journal of Dairy Science* 63, 283-289.

Laster, D.B., Glimp, H.A. and Gregory, K.E. (1973) Effects of early weaning on postpartum reproduction of cows. *Journal of Animal Science* 36, 734-740.

Leaver, J.D. (1977) Rearing of dairy cattle. 7. Effect of level of nutrition and body condition on the fertility of heifers. *Animal Production* 25, 219-224.

Legge, M. and Sellens, M.M. (1991) Free radical scavengers ameliorate the 2-cell block in mouse embryo culture. *Human Reproduction* 6, 867-871.

Leng, R.A. (1970) Glucose synthesis in ruminants. *Advances in Veterinary Science* 14, 209-260.

Leonard, M.C., Buttery, P.J. and Lewis, D. (1977) The effects on glucose metabolism of feeding a high-urea diet to sheep. *British Journal of Nutrition* 38, 455-462.

Liebholz, J. and Kellaway, R.C. (1984) The utilization of low quality roughages. 1. The role of nitrogen and energy supplements. *Australian Meat Research Committee Review* 48, 1-21.

Liggins G.C. (1982) The fetus and birth. In: Austin, C.R. and Short, R.V. (eds) *Reproduction in Mammals, Book 2. Embryonic and Fetal Development*, 2nd edn. Cambridge University Press, Cambridge, p. 115.

Lindsay, D.B. and Setchell, B.P. (1976) The oxidation of glucose, ketone bodies and acetate by the brain of normal and ketonaemic sheep. *Journal of Physiology* 259, 801-823.

Little, D.A. (1968) Effect of dietary phosphate on the voluntary consumption of Townsville lucerne (*Stylosanthes humilis*) by cattle. *Proceedings of the Australian Society of Animal Production* 7, 376-380.

Little, D.A. (1970) Factors of importance in the phosphorus nutrition of beef cattle in northern Australia. *Australian Veterinary Journal* 46, 241-248.

Long, T.A., Tillman, A.D., Nelson, A.B., Gallup, W.D. and Davis, B. (1957) Availability of phosphorus in mineral supplements for beef cattle. *Journal of Animal Science* 16, 444-450.

Lotan, E. and Adler, J.H. (1960) Early effects of excessive alfalfa feeding in bovine fertility. *Refuah Veterinarith* 23, 112-110.

Lotan, E., Ziv, E., Levy, E., Marton, M. and Adler, J.H. (1988) Experimental manipulation of post partum energy partition in high yielding dairy cows. *Israeli Journal of Veterinary Medicine* 44, 159-167.

Lotthammer, K.H., Cooke, B.C. and Friesecke, H. (1978) Importance of beta-carotene for bovine fertility. In: *Roche Symposium*, London, Roche, Switzerland.

Lucy, M.C., Staples, C.R., Michel, F.M., Thatcher, W.W. and Bolt, D.J. (1991) Effect of feeding calcium soaps to early postpartum dairy cows on plasma prostaglandin $F_{2\alpha}$, luteinizing hormone, and follicular growth. *Journal of Dairy Science* 74, 483-489.

Lucy, M.C., Beck, J., Staples, C.R., Head, H.H., Sota, R.L.D. and Thatcher, W.W.

(1992a) Follicular dynamics, plasma metabolites, hormone and insulin-like growth factor (IGF-1) in lactating cows with a positive or negative energy balance during the preovulatory period. *Reproduction, Nutrition, Development* 32, 331–341.

Lucy, M.C., Staples, C.R., Thatcher, W.W., Erickson, P.S., Cleale, R.M., Firkins, L. and Clark, J.H. (1992b) Influence of diet, dry matter intake, milk production and energy balance on time of post-partum ovulation and fertility in dairy cows. *Animal Production* 54, 323–331.

Lusby, K.S. and Wettemann, R.P. (1980) Effects of early weaning calves from first calf heifers on calf and heifer performance. *Agricultural Research Station Reports, Oklahoma State University Miscellaneous Publication* 107, 55–58.

McCann, J.P. and Hansel, W. (1986) Relationships between insulin and glucose metabolism and pituitary–ovarian functions in fasted heifers. *Biology of Reproduction* 34, 630–641.

McClary, D., McGuffey, R.K. and Green, H.B. (1990) Bovine somatotropin: Part 2. *Agri-Practice* 11, 5–11.

McClure, T.J. (1961) An apparent nutritional lactational stress infertility in dairy herds. *New Zealand Veterinary Journal* 9, 107–12.

McClure, T.J. (1965a) Experimental evidence for the occurrence of nutritional infertility in otherwise clinically healthy pasture-fed lactating dairy cows. *Research in Veterinary Science* 6, 202–208.

McClure, T.J. (1965b) A nutritional cause of low non-return rates in dairy herds. *Australian Veterinary Journal* 41, 119–122.

McClure, T.J. (1966) Infertility in mice caused by fasting at about the time of mating. I. Mating behaviour and littering rates. *Journal of Reproduction and Fertility* 12, 243–248.

McClure, T.J. (1967a) Infertility in mice caused by fasting at about the time of mating. II. Pathological changes. *Journal of Reproduction and Fertility* 13, 387–391.

McClure, T.J. (1967b) Infertility in mice caused by fasting at about the time of mating. III. Pathogenesis. *Journal of Reproduction and Fertility* 13, 393–403.

McClure, T.J. (1967c) The effect of environmental stresses on early pregnancy with particular reference to nutritional stress in cattle and mice. PhD Thesis, The University of Sydney.

McClure, T.J. (1968) Hypoglycaemia, an apparent cause of infertility of lactating cows. *British Veterinary Journal* 124, 126–130.

McClure, T.J. (1970) An experimental study of the causes of a nutritional and lactational stress infertility of pasture-fed cows, associated with loss of bodyweight at about the time of mating. *Research in Veterinary Science* 11, 247–254.

McClure, T.J. (1977a) Effects of food intake and composition on the concentration of glucose in the blood of lactating cattle. *Australian Journal of Agricultural Research* 28, 333–339.

McClure, T.J. (1977b) Effect of feed quality and stage of lactation on the concentration of glucose in the blood of lactating cattle. *Australian Journal of Agricultural Research* 28, 341–344.

McClure, T.J. and Payne, J.M. (1978) Observations on the first service non-return

rates of hypoglycaemic, concentrate-fed dairy herds. *Australian Veterinary Journal* 54, 7-9.
McClure, T.J. and Saunders, J. (1985) Effects of withholding food for 0-72 h on mating, pregnancy rate and pituitary function in female rats. *Journal of Reproduction and Fertility* 74, 57-64.
McClure, T.J., Nancarrow, C.D. and Radford H.M. (1978) The effect of 2-deoxy-D-glucose on ovarian function of cattle. *Australian Journal of Biological Science* 31, 183-186.
McClure, T.J., Eamens, G.J. and Healy, P.J. (1986) Improved fertility in dairy cows after treatment with selenium pellets. *Australian Veterinary Journal* 63, 144-146.
McDonald, P., Edwards, R.A. and Greehalgh, J.F.D. (1991) *Animal Nutrition*, 4th edn. Longmans Scientific and Technical, Harlow, UK.
McDonald, R.J., McKay, G.W. and Thomson, J.D. (1962) The use of organic iodine in the treatment of repeat breeder cows. In: *Proceedings of the 4th International Congress on Animal Reproduction* No. 3, pp. 679-681.
McGhie, T.K. (1991) Analysis of serum methylmalonic acid for the determination of cobalt deficiency in cattle. *Journal of Chromatography, Biomedical Applications* 566, 215-222.
McGowan, A.A. and Earle, D.F. (1978) A rising plate meter for assessing pasture yield. *Proceedings of the Australian Society of Animal Production* 12, 211.
McGowan, M.R. (1991) Studies on early embryonic mortality in cattle. PhD Thesis, The University of Sydney.
Machlin, L.J. (1984) *Handbook of Vitamins*. Marcel Dekker, NY.
McLaren, A. (1982) The embryo. In: Austin, C.R. and Short, R.V. (eds) *Reproducton in Mammals. Book 2. Embryonic and Fetal Development*, 2nd edn. Cambridge University Press, Cambridge, pp. 1-25.
McLeod, B.J. and Williams, B.E. (1991) Incidence of ovarian dysfunction in post partum dairy cows and the effectiveness of its clinical diagnosis and treatment. *Veterinary Record* 128, 121-134.
McMeekan, C.P. (1956) Grazing management and animal production. In: *Proceedings of the 7th International Grassland Conference*, Session 3, pp. 146-156.
McMeekan, C.P. and Walshe, M.J. (1963) The inter-relationships of grazing method and stocking rate in the efficiency of pasture utilization by dairy cattle. *Journal of Agricultural Science* 61, 147-166.
Macmillan, K.L., Taufa, V.K., Day, A.M. and Peterson, A.J. (1991) Effects of supplemental progesterone on pregnancy rates in cattle. *Journal of Reproduction and Fertility*, Supplement 43, 304.
MacPherson, A. (1985) Long-term effectiveness of copper oxide wire treatment in the bovine. *Trace Elements in Man and Animals* 5, 733-735.
MacPherson, A., Moon, F.E. and Voss, R.C. (1973) Some effects of feeding young steers on a diet deficient in both cobalt and copper. *British Veterinary Journal* 129, 414-425.
McQueen, J.K. and Fink, G. (1988) Changes in local cerebral glucose utilisation associated with the spontaneous ovulatory surge of luteinizing hormone in the rat. *Neuroendocrinology* 47, 551-555.

McTaggart, H.S. (1959) "Milk lameness": an aphosphorosis of heavy-milking cows. *Veterinary Record* 71, 709–714.

MAFF (Ministry of Agriculture, Fisheries and Food, Department of Agriculture and Fisheries for Scotland, Department of Agriculture for Northern Ireland) (1984) *Energy Allowances and Feeding Systems for Ruminants*, Reference book 433. HMSO, London.

MAFF (1992) *UK Tables of Feed Composition and Nutritive Value for Ruminants*, 2nd edn. Ministry of Agriculture, Fisheries and Food, UK.

Makeham, J.F. and Malcolm, L.R. (1983) *The Farming Game*. Gill, Armidale, New South Wales, p. 184.

Mannetje, L't. (1978) *Measurements of Grassland Vegetation and Animal Production*. Commonwealth Agricultural Bureaux, Farnham Royal, UK.

Mannetje, L't. (1982) Problems of animal production from tropical pastures. In: Hacker, J.B. (ed.) *Nutritional Limitations to Animal Production from Pastures*. Commonwealth Agricultural Bureaux, Farnham Royal, UK, pp. 67–83.

Marston, H.R., Allen, S.H. and Smith, R.M. (1961) Primary metabolic defect supervening on vitamin B_{12} deficiency in the sheep. *Nature* 190, 1085–1088.

Mason, G.L. and Randel, R.D. (1983) Effect of monensin and suckling on the GnRH induced luteinizing hormone surge and the effect of monensin on the postpartum interval in Brangus cows. *Theriogenology* 19, 331–342.

Mason, K.E. (1979) A conspectus of research on copper metabolism and requirements of man. *Journal of Nutrition* 109, 1979–2066.

Mazur, M. and YoungLai, E.V. (1986) Glucose metabolism in perinatal gonads of the rabbit. *Journal of Reproduction and Fertility* 77, 207–209.

Mee, M.O., Stevenson, J.S., Alexander, B.M. and Sasser, R.G. (1993) Administration of GnRH influences pregnancy rates, serum concentrations of LH, FSH, estradiol-17β, pregnancy-specific protein B, and progesterone, proportion of luteal cell types and *in vitro* production of progesterone in dairy cows. *Journal of Animal Science* 71, 185–198.

Meijs, J.A.C. (1981) Herbage intake by grazing dairy cows. In: *Agricultural Research Report*. Pudoc, Wageningen.

Meites, J. (1953) Relation of nutrition to endocrine–reproductive functions. *Iowa State College Journal of Science* 28, 19–44.

Meites, J. (1984) Effects of opiates on neuroendocrine functions in animals: overview. In: Delitala, G. *et al.* (eds) *Opioid Modulation of Endocrine Function*. Raven, NY, pp. 53–63.

Mills, C.F., Dalgarno, A.C. and Wenham, G. (1976) Biochemical and pathological changes in tissues of Friesian cattle during the experimental induction of copper deficiency. *British Journal of Nutrition* 35, 309–330.

Miltimore, J.E., van Ryswyk, A.L., Pringle, W.L., Chapman, F.M. and Kalnin, C.M. (1975) Selenium concentrations in British Columbia forages, grains and processed feeds. *Canadian Journal of Animal Science* 55, 101–111.

Minson, D.J. (1982) Effects of chemical and physical composition of herbage eaten upon intake. In: Hacker, J.B. (ed.) *Nutritional Limits to Animal Production from Pastures*. Commonwealth Agricultural Bureaux, Farnham Royal, UK, pp. 167–182.

Minson, D.J. (1990) *Forage in Ruminant Nutrition*. Academic Press, London.

Minson, D.J. and McLeod, M.N. (1970) The digestibility of temperate and tropical grasses. In: *Proceedings of the 11th International Grasslands Conference.* Surfers' Paradise, Queensland, pp. 719–722.

Moller, K. and Fielden, E.D. (1981) Pre-mating injection of an analogue of gonadotrophin-releasing hormone (GnRH) and pregnancy rates to first insemination. *New Zealand Veterinary Journal* 29, 214–215.

Moller, K. and Shannon, P. (1972) Body weight change and fertility of dairy cows. *New Zealand Veterinary Journal* 20, 47–48.

Montgomery, M.J. and Baumgardt, B.R. (1965a) Regulation of food intake in ruminants. 1. Pelleted rations varying in energy concentration. *Journal of Dairy Science* 48, 569–574.

Montgomery, M.J. and Baumgardt, B.R. (1965b) Regulation of food intake in ruminants. 2. Rations varying in energy concentration and physical form. *Journal of Dairy Science* 48, 1623–1628.

Morgan, D.E. and Trinder, H. (1980) The composition and nutritional value of some tropical and sub-tropical by-products. In: *By-Products and Wastes in Animal Feeding. Occasional Publication No. 3.* British Society of Animal Production, Reading, pp. 91–111.

Morgan, W.F. and Lean, I.J. (1993) Gonadotrophin-releasing hormone treatment in cattle: a meta-analysis of the effects on conception at the time of insemination. *Australian Veterinary* 70, 205–209.

Morley, F.H.W., Axelsen, A. and Cunningham, R.B. (1976) Liveweight at joining and fertility in beef cattle. *Proceedings of the Australian Society of Animal Production* 11, 201–204.

Morrow, D., Thomas, J.W. and Main, R.J. (1981) Effects of vitamin E and selenium on periparturient diseases and fertility in dairy cattle. *Bovine Practitioner* 16, 80–84.

Moss, G.E., Parfet, J.R, Marvin, C.A., Allrich, R.D. and Diekman, M.A. (1985) Pituitary concentrations of gonadotrophins and receptors for GnRH in suckled beef cows at various intervals after calving. *Journal of Animal Science* 60, 285–293.

Munro, I.B. (1957) Infectious and non-infectious herd infertility in East Anglia. *Veterinary Record* 69, 125–129.

Munro, I.B. (1975) Copper deficiency in cattle. In: *Copper Farming Symposium.* Royal Zoological Society, London and Copper Development Association, Potters Bar, UK, pp. 63–67.

Mylrea, P.J. and Bayfield, R.F. (1968) Concentrations of some components in the blood and serum of apparently healthy dairy cattle. 1. Electrolytes and minerals. *Australian Veterinary Journal* 44, 565–569.

Nagatsu, T. (1973) *Biochemistry of Catecholamines.* University Park Press, Baltimore.

Nelson, B.D., Ellzey, H.D., Morgan, E.B. and Allen, M. (1968) Effects of feeding lactating dairy cows varying forage-to-concentrate ratios. *Journal of Dairy Science* 51, 1796–1800.

Nett, T.M. (1987) Function of the hypothalamic–hypophysial axis during the post partum period in ewes and cows. *Journal of Reproduction and Fertility*, Supplement 34, 201–213.

Nolan, C.J., Bull, R.C., Sasser, R.G., Ruder, C.A., Panlasigui, P.M., Schoeneman, H.M. and Reeves, J.J. (1988) Postpartum reproduction in protein restricted beef cows: effect on the hypothalamic–pituitary–ovarian axis. *Journal of Animal Science* 66, 3208–3217.

Norman, M.J.T. (1963) Dry season protein and energy supplements for beef cattle on native pastures at Katherine, N.T. *Australian Journal of Experimental Agriculture and Animal Husbandry* 3, 280–283.

Norton, J.H. and Hogan, J.P. (1993) Lack of association between abortion and blood ammonia and methaemoglobin concentrations in dairy cows grazing improved pastures on the Atherton Tablelands. *Australian Veterinary Journal* 70, 194–195.

NRC (1978) *Nutrient Requirements of Domestic Animals*. National Academy of Sciences, Washington, DC.

NRC (1982) *United States – Canadian Tables of Feed Compostion*. National Academy Press, Washington, DC.

NRC (1984) *Nutrient Requirements of Beef Cattle*. National Academy Press, Washington, DC.

NZDB (1961) *Farm Production Report No. 37*. New Zealand Dairy Board, Wellington, NZ.

O'Connor, M.B. (1982) The provision of summer/autumn feed for the dairy herd. In: Macmillan, K.L. and Taufa, V.K. (eds) *Proceedings of the Conference on Dairy Production from Pasture*. New Zealand and Australian Societies of Animal Production, Ruakura Animal Research Station, Hamilton, New Zealand, pp. 367–374.

Ohba, C., Ichijo, S. and Osame, S. (1992) Selenium and tocopherol levels in the serum and organs of aborted and premature fetuses and calves that died just after birth. *Journal of the Japan Veterinary Medical Association* 45, 476–479.

O'Moore, L.B. (1950) Observations on phosphorus deficiency in the grazing cow. *Irish Veterinary Journal* 4, 198–209 and 218–227.

Orskov, E.R. (1982) *Protein Nutrition in Ruminants*. Academic Press, London.

Ostrowski-Meissner, H.T. (1990) *Australian Feed Composition Tables*. Australian Feed Information Centre, CSIRO, Blacktown, NSW.

Oxenreider, S.L. (1968) Effects of suckling and ovarian function on postpartum reproductive activity in beef cows. *American Journal of Veterinary Research* 29, 2099–2102.

Oxenreider, S.L. and Wagner, W.C. (1971) Effect of lactation and energy intake on postpartum ovarian activity in the cow. *Journal of Animal Science* 33, 1026–1031.

Palmer, L.S., Gullickson, T.W., Boyd, W.L., Fitch, C.P. and Nelson, J.W. (1941) The effect of rations deficient in phosphorus and protein on ovulation, estrous and reproduction of dairy heifers. *Journal of Dairy Science* 24, 199–210.

Paquay, R., Godeau, J.M., de Baere, R. and Lousse, A. (1973a) The effects of the protein content of the diet on the performance of lactating cows. *Journal of Dairy Research* 40, 93–103.

Paquay, R., Godeau, J.M., de Baere, R. and Lousse, A. (1973b) Utilization of nutrients by the dairy cow and optimal N : energy ratio in the diet. *Journal of Dairy Research* 40, 329–337.

Parker, O.F. (1966) Supplementary feeding of hay to dairy cows after calving. In: *Proceedings of the Ruakura Farmers' Week Conference.* New Zealand Department of Agriculture, Wellington, pp. 159–167.

Parker, O.F. (1973) Precision in field budgeting in practice. *Proceedings of the New Zealand Grassland Association* 35, 127–134.

Pawlina, E., Jaczewski, S. and Monkiewicz, J. (1988) [Length of life, performance and reasons for culling of cows in an area contaminated by a copper smelter.] *Zootecknika* 30, 87–94 (in Polish).

Payne, J.M. (1977) *Metabolic Diseases in Farm Animals.* Heinemann, London.

Payne, J.M., Rowlands, G.J., Manston, R. and Dew, S.M. (1973) A statistical appraisal of the results of metabolic profile tests on 75 dairy herds. *British Veterinary Journal* 129, 370–381.

Perry, R.C., Corah, L.R., Kiracofe, G.H., Stevenson, J.S. and Beal, W.E. (1991) Endocrine changes and ultrasonography of ovaries in suckled beef cows during resumption of postpartum estrous cycles. *Journal of Animal Science* 69, 2548–2555.

Peters, A.R., Lamming, G.E. and Fisher, M.W. (1981) A comparison of plasma LH concentrations in milked and suckling post-partum cows. *Journal of Reproduction and Fertility* 62, 567–573.

Peterson, R.R., Webster, R.C., Rayner, B. and Young, W.C. (1952) The thyroid and reproductive performance in the adult female guinea pig. *Endocrinology* 51, 504–518.

Phillippo, M., Humphries, W.R., Lawrence, C.B. and Price, J. (1982) Investigation of the effect of copper status and therapy on fertility in beef suckler herds. *Journal of Agricultural Science (Cambridge)* 99, 359–364.

Phillippo, M., Humphries, W.R. and Atkinson, T. (1985) The effect of molybdenum on fertility in the cow. *Proceedings of the Nutrition Society* 44, 82A.

Pike, I.H. and Tatterson, I.N. (1980) The utilization of fish by-products and waste in animal feeding. In: *By-Products and Wastes in Animal Feeding. Occasional Publication No. 3.* British Society of Animal Production, Reading, pp. 85–90.

Playne, M.J. (1969) Effects of sodium sulphate and gluten supplements on the intake and digestibility of a mixture of spear grass and Townsville lucerne hay by sheep. *Australian Journal of Experimental Agriculture and Animal Husbandry* 9, 393–399.

Poff, J.P., Fairchild, D.L. and Condon, W.A. (1988) Effects of antibiotics and medium supplements on steroidogenesis in cultured cow luteal cells. *Journal of Reproduction and Fertility* 82, 135–143.

Potter, E.L., Cooley, C.O., Richardson, L.F., Raun, A.P. and Rathmacher, R.P. (1976) Effect of monensin on performance of cattle fed forage. *Journal of Animal Science* 43, 665–669.

Putnam, D.N. and Henderson, H.O. (1946) The effect of pregnancy on the body weight of dairy cows. *Journal of Dairy Science* 29, 657–661.

Putnam, M.E. and Comben, N. (1987) Vitamin E. *Veterinary Record* 121, 541–545.

Rae, D.O., Kunkle, W.E., Chenoweth, P.J., Sand, R.S. and Tran, T. (1993) Relationship of parity and body condition score to pregnancy rates in Florida beef cattle. *Theriogenology* 39, 1143–1152.

Rammell, C.G. (1983) Vitamin E status of cattle and sheep. 1. A background review. *New Zealand Veterinary Journal* 31, 179–181.

Randel, R.D. (1990) Nutrition and postpartum rebreeding in cattle. *Journal of Animal Science* 68, 853–862.

Randel, R.D., Rutter, L.M. and Rhodes, R.C (1982a) III. Monensin effects on the estrogen induced LH surge in pre-puberal heifers. In: *Beef Cattle Research in Texas*. Texas Agricultural Research Station, College Station, Texas, pp. 20–23.

Randel, R.D., Rutter, L.M. and Rhodes R.C. (1982b) Effect of monensin on the estrogen induced LH surge in pre-puberal heifers. *Journal of Animal Science* 54, 806–810.

Rasbech, N.O. (1968) [Manganese concentrations in the ovaries of heifers fed various amounts of manganese. *Zuchthygiene* 3, 57–62 (in German).

Raymond, W.F. (1964) The efficient use of grass. *Journal of the British Grassland Society* 19, 81–89.

Rega, A.F. and Garrahan, P.J. (1986) *The Ca^{2+} Pump of Plasma Membranes*. CRC, Boca Raton, Florida.

Reid, I.C. and McQueen, I.P. (1985) Cobalt supplementation of beef cattle. *Trace Elements in Man and Animals* 5, 739–741.

Reid, I.M. (1980) Incidence and severity of fatty liver in dairy cows. *Veterinary Record* 107, 281–284.

Reid, I.M. (1983) Reproductive performance and fatty liver in Guernsey cows. *Animal Reproduction Science* 5, 275–279.

Reid, J.T., Loosli, J.K., Trimberger, G.W., Turk, K.L., Asdell, S.A. and Smith, S.E. (1964) Causes and prevention of reproductive failures in dairy cattle. 4. Effect of plane of nutrition during early life on growth, reproduction, health and longevity of Holstein cows. (I) Birth to fifth calving. *Cornell University Agricultural Experimental Station, Bulletin* No. 987.

Reisen, J.W., Saiduddin, S., Tyler, W.J. and Casida, L.E. (1968) Relation of post partum interval to corpus luteum development, pituitary prolactin activity and uterine involution in dairy cows (effect of suckling). *Wisconsin Agricultural Experimental Station Research Bulletin* No. 270.

Reynolds, W.L., DeRouen, T.M., High, J.W., Wiltbank, J.N., Warwick, E.J. and Temple, R.S. (1964) Evaluation of pastures in terms of reproduction of beef cattle. *Journal of Animal Science* 25, 890.

Richards, M.W., Wettemann, R.P. and Schoenemann, H.M. (1989a) Nutritional anestrus in beef cows: body weight change, body condition, luteinizing hormone in serum and ovarian activity. *Journal of Animal Science* 67, 1520–1526.

Richards, M.W., Wettemann, R.P. and Shoenemann, H.M. (1989b) Nutritional anestrus in beef cows: concentrations of glucose and nonesterified fatty acids in plasma and insulin in serum. *Journal of Animal Science* 67, 2354–2362.

Roberts, R.M., Klemann, S.W., Leaman, D.W., Bixby, J.A., Cross, J.C., Farin, C.W., Imakawa, K. and Hansen, T.R. (1991) The polypeptides and genes for ovine and bovine trophoblast protein-1. *Journal of Reproduction and Fertility*, Supplement 43, 3–12.

Robinson, D.W. and Sageman, R. (1967) The nutritive value of some pasture species in north-western Australia during the late dry season. *Australian Journal of Experimental Agriculture and Husbandry* 7, 533–539.

Rogers, G.L., Bryant, A.M., Jury, K.E. and Hutton, J.B. (1979) Silage and dairy cow production. 1. Digestible energy intake and yield and composition of milk of cows fed pasture and pasture silages. *New Zealand Journal of Agricultural Research* 22, 511–522.

Rogers, G.L., Porter, R.H.D. and Robinson, I. (1982) Comparison of perennial ryegrass and white clover for milk production. In: Macmillan, K.L. and Taufa, V.K. (eds) *Proceedings of the Conference on Dairy Production from Pasture.* New Zealand and Australian Societies of Animal Production, Ruakura Animal Research Station, Hamilton, New Zealand, pp. 213–214.

Rojas, M.A., Dyer, I.A. and Cassatt, W.A. (1965) Manganese deficiency in the bovine. *Journal of Animal Science* 24, 664–667.

Rook, J.A.F. and Line, C. (1961) The effect of the plane of energy nutrition of the cow on the secretion in milk of the constituents of the solids-not-fat fraction and on the concentrations of certain blood-plasma consitutents. *British Journal of Nutrition* 15, 109–119.

Ropstad, E. (1988) Constituents of blood and milk in relation to fertility, nutrition and metabolic status in dairy cows. Thesis, Norwegian College of Veterinary Medicine, Oslo.

Roy, J.H.B. (1980) *The Calf*, 4th edn. Butterworths, London.

Rutter, L.M. and Manns, J.G. (1987) Hypoglycaemia alters pulsatile luteinizing hormone secretion in the postpartum beef cow. *Journal of Animal Science* 64, 479–488.

Rutter, L.M. and Manns, J.G. (1988) Follicular phase gonadotropin secretion in cyclic postpartum beef cows with phlorizin-induced hypoglycaemia. *Journal of Animal Science* 66, 1194–1200.

Rutter, L.M. and Manns, J.G. (1991) Insulin-like growth factor 1 in follicular development and function in postpartum beef cows. *Journal of Animal Science* 69, 1140–1146.

Rutter, L.M. and Randel, R.D. (1984) Postpartum nutrient intake and body condition: effect on pituitary function and onset of estrus in beef cattle. *Journal of Animal Science* 58, 265–274.

Rutter, L.M., Randel, R.D., Schelling, G.T. and Forest, D.W. (1983) Effect of abomasal infusion of propionate on the GnRH-induced luteinizing hormone release in prepuberal heifers. *Journal of Animal Science* 56, 1167–1173.

Ryan, D.P., Rodriguez, H.F., Thompson, D.L., Saxton, A.M. and Godke, R.A. (1992) Luteal maintenance in cattle after conceptus death during the first trimester of gestation. *Journal of Animal Science* 70, 836–840.

Ryot, K.D., Sharma, B.K. and Panwar, D.D. (1990) Effect of iodine therapy in anoestrous bovines. *Indian Journal of Animal Reproduction* 11, 144–145.

Saiko, A.A. and Krukovets, M.K. (1976) [The effect of methionine supplements on reproductive function of cows.] *Zhivotnovodstvo* 12, 62–64 (in Russian).

Saito, A. (1981) Changes in hypothalamic LH-RH content induced by estradiol injection into female rats. *Acta Obstetria et Gynaecologica Japonica* 33, 1227–1234.

Sasser, R.G., Williams, R.J., Bull, R.C., Ruder, C.A. and Falk, D.G. (1988) Postpartum reproductive performance in crude protein-restricted beef cows: return to estrus and conception. *Journal of Animal Science* 66, 3033–3039.

Scales, G.H. and Stevenson, J.R. (1976) Nutrition of the beef breeding cow. In: *Proceedings of the Ruakura Farmers' Week Conference*. New Zealand Department of Agriculture, Wellington, pp. 45-48.

Schultz, L.H. (1969) Relationship of rearing rate of dairy heifers to mature performance. *Journal of Dairy Science* 52, 1321-1329.

Scott, J.D.J. and Smeaton, D.C. (1980) In: Scott, J.D.J., Lamont, N., Smeaton, D.C. and Hudson, S.J. (eds) *Sheep and Cattle Nutrition*. NZ Ministry of Agriculture, Wellington, p. 26.

Segerson, E.C., Murray, F.A., Moxon, A.L., Redman, D.R. and Conrad, H.R. (1977) Selenium/vitamin E: role in fertilization of bovine ova. *Journal of Dairy Science* 60, 1001-1005.

Sen, K.K., Azhar, S. and Menon, K.M.J. (1979) Evidence for the involvement of an energy-dependent process in gonadotropin-releasing hormone-stimulated luteinizing hormone release by rat anterior pituitary. *Endocrinology* 105, 1158-1161.

Setchell, B.P., Bassett, J.M., Hinks, N.T. and Graham N.McC. (1972) The importance of glucose in the oxidative metabolism of the pregnant uterus and its contents in conscious sheep with some preliminary observations on the oxidation of fructose and glucose by fetal sheep. *Quarterly Journal of Experimental Physiology* 57, 257-266.

Short, R.E. and Adams, D.C. (1988) Nutritional and hormonal interrelationships in beef cattle reproduction. *Canadian Journal of Animal Science* 68, 29-39.

Shrick, F.N., Spitzer, J.C., Jenkins, T.C., Henricks, D.M. and Althen, T.G. (1990) Effect of dietary energy restriction on metabolic and endocrine responses during the estrous cycle of the suckled beef cow. *Journal of Animal Science* 68, 3313-3321.

Siebert, B.D., Playne, M.J. and Edye, L.A. (1975) The effects of climate and nutrient supplementation on the fertility of heifers in North Queensland. *Proceedings of the Australian Society of Animal Production* 11, 249-252.

Simkins, K.L., Suttie, J.W. and Baumgardt, B.R. (1965a) Regulation of food intake in ruminants. 3. Variation in blood and rumen metabolites in relation to food intake. *Journal of Diary Science* 48, 1629-1634.

Simkins, K.L., Suttie, J.W. and Baumgardt, B.R. (1965b) Regulation of food intake in ruminants. 4. Effect of acetate, propionate, butyrate and glucose on voluntary food intake in dairy cattle. *Journal of Dairy Science* 48, 1635-1642.

Simonik, I. and Pavelka, J. (1989) [Residues of toxic elements in the cervical mucus of cows in relation to conception.] *Veterinarstvi* 39, 107-108 (in Czechoslovakian).

Singer, D.D. (1980) By-products from the food and dairy industries. In: *By-products and Wastes in Animal Feeding. Occasional Publication No. 3*. British Society of Animal Production, Reading, pp. 71-78.

Sklan, D., Moallem, U. and Folman, Y. (1991) Effect of feeding calcium soaps of fatty acids on production and reproductive responses in high producing lactating cows. *Journal of Dairy Science* 74, 510-517.

Smeaton, D.C., McCall, D.G. and Wadams, T.K. (1983) Effects of pasture allowance level after calving on performance of beef cows on hill country. *New Zealand Journal of Experimental Agriculture* 11, 303-308.

Smith, D. (1960) Yield and chemical composition of oats for forage with advance in maturity. *Agronomy Journal* 52, 637–639.

Smith, J.F., Payne, E., Tervit, H.R., McGowan, L.T., Fairclough, R., Kilgour, R. and Goold, P.G. (1981) The effect of suckling upon the endocrine changes associated with anoestrus in identical twin dairy cows. *Journal of Reproduction and Fertility*, Supplement 30, 241–249.

Smith, M.F., Burrell, W.C., Shipp, L.D., Sprott, L.R., Songster, W.N. and Wiltbank, J.N. (1979) Hormone treatments and use of calf removal in postpartum beef cows. *Journal of Animal Science* 48, 1285–1294.

Sommer, H. (1975) Preventive medicine in dairy cows. *Veterinary Medical Review* 1/2, 42–63.

Sommerville, S.H., Lowman, B.G. and Deas, D.W. (1979) The effect of plane of nutrition during lactation on the reproductive performance of beef cows. *Veterinary Record* 104, 95–97.

Sorensen, A.M., Hansel, W., Hough, W.H., Armstrong, D.T., McEntee, K. and Bratton, W.R. (1959) Causes and prevention of reproductive failures in dairy cattle. 1. Influence of underfeeding and overfeeding on growth and development of Holstein heifers. *Cornell University Agricultural Experimental Station Bulletin* No. 936.

Southcott, W.H., Wheeler, J.L., Hill, M.K. and Hedges, D.A. (1972) Effect of subdivision, stocking rate, anthelmintic and selenium on the productivity of Hereford heifers. *Proceedings of the Australian Society of Animal Production* 9, 408–411.

Spiekers, H. and Pfeffer, E. (1990) Studies of the effect of supplementing protected methionine (HMM-Ca) to dairy cows on milk yield and fertility. *Archives of Animal Nutrition* 40, 449–458.

Sprangers, S.A. and Piacsek, B.E. (1988) Increased suppression of luteinizing hormone secretion by chronic and acute estradiol administration in underfed adult female rats. *Biology of Reproduction* 39, 81–87.

Sprott, L.R., Goehring, T.B., Beverly, J.R. and Corah, L.R. (1988) Effects of ionophores on cow herd production: a review. *Journal of Animal Science* 66, 1340–1346.

Stabenfeldt, G.H., Ewing, L.L. and McDonald, L.E. (1969) Peripheral plasma progesterone levels during the bovine oestrous cycle. *Journal of Reproduction and Fertility* 19, 433–442.

Stoikov, D. (1978) [Methionine and infertility in cows.] *Veterinarna Sbirka* 76, 30–32 (in Bulgarian).

Storry, J.E. and Rook, J.A.F. (1966) The relationship in the cow between milk-fat secretion and ruminal volatile fatty acids. *British Journal of Nutrition* 20, 217–228.

Stowe, H.D., Thomas, J.W., Johnson, T., Marteniuk, J.V., Morrow, D.A. and Ullrey, D.E. (1988) Responses of dairy cattle to long-term and short-term supplementation with oral selenium and vitamin E. *Journal of Dairy Science* 71, 1830–1839.

Suttle, N.F. (1981) Effectiveness of orally administered cupric oxide needles in alleviating hypocupraemia in sheep and cattle. *Veterinary Record* 108, 417–420.

Sutton, J.D., Hart, I.C., Morant, S.V., Schuller, E. and Simmonds, A.D. (1988)

Frequency of feeding for lactating cows: diurnal patterns of hormones and metabolites in peripheral blood in relation to milk-fat concentration. *British Journal of Nutrition* 60, 265–274.

Suzuki, O., Sato, M. and Kubota, Y. (1982) [Effects of underfeeding on estrous behaviour and ovarian function in postpubertal heifers.] *Japanese Journal of Animal Reproduction* 28, 205–210 (in Japanese).

Swanson, E.W. and Hinton, S.A. (1962) Effects of adding concentrates to ad libitum roughage feeding in the dry period. *Journal of Dairy Science* 45, 48–54.

Symonds, H.W. and Baird, G.D. (1975) Evidence for the absorption of reducing sugar from the small intestine of the dairy cow. *British Veterinary Journal* 131, 17–22.

Tamminga, S. and von Hellemond, K.K. (1977) The protein requirements of dairy cattle and developments in the use of protein, essential amino acids and non-protein nitrogen in the feeding of cattle. In: *Nitrogen and Non-protein Nitrogen for Ruminants.* Pergamon, Oxford, pp. 9–31.

Taylor, R.F., Puls, R. and McDonald, K.R. (1979) Bovine abortions associated with selenium deficiency in western Canada. *Proceedings of the American Association of Veterinary Laboratory Diagnosticians* 22, 77–84.

Teleni, E., Rowe, J.B. and Croker, K.P. (1984) Increased ovulation rates in ewes given intravenous infusion of energy-yielding substrates. *Proceedings of the Nutrition Society of Australia* 9, 158.

Teleni, E., Rowe, J.B., King, W.R., Murray, P.J. and Croker, K.P. (1985) The effect of intravenous infusions of either glucose, acetate or glucose and acetate on ovulation rates in ewes. *Proceedings of the Nutrition Society of Australia* 10, 195.

Teleni, E., Rowe, J.B., Croker, K.P., Murray, P.J. and King, W.R. (1989) Lupins and energy-yielding nutrients in ewes. II. Responses in ovulation rate in ewes to increased availability of glucose acetate and amino acids. *Reproduction Fertility and Development* 1, 117–125.

Tervit, H.R., Smith, J.F., Goold, P.G., Jones, K.R. and Vandien, J.J.D. (1982) Reproductive performance of beef cows following temporary removal of calves. *Proceedings of the New Zealand Society of Animal Production* 42, 83–85.

Thain, R.I. (1967) Evidence for the widespread involvement of clover pastures in bovine infertility in Tasmania. *Australian Journal of Science* 29, 220–221.

Theiler, A. and Green, H.H. (1932) Aphosphorosis in ruminants. *Nutrition Abstracts and Reviews* 1, 359–385.

Thibier, M., Chapalgaonkar, K., Joshi, A., Karbade, V. and Recca, A. (1983) Use of heat detection paste in dairy cattle in France. *Veterinary Record* 113, 128–130.

Thomson, D.J., Beever, D.E., Haines, M.J., Cammell, S.B., Evans, R.T., Dhanoa, M. and Austin, A.R. (1985) Yield and composition of milk from Friesian cows grazing either perennial ryegrass or white clover in early lactation. *Journal of Dairy Research* 52, 17–31.

Treacher, R.J., Reid, I.M. and Roberts, C.J. (1986) Effect of body conditon at calving on the health and performance of dairy cows. *Animal Production* 43, 1–6.

Trinder, N., Woodhouse, C.D. and Renton, C.P. (1969) The effect of vitamin E and selenium on the incidence of retained placentae in dairy cows. *Veterinary Record* 85, 550–553.

Tuff, P. (1923) Osteomalacia and its occurrences in cattle in Norway. *Journal of Comparative Pathology and Therapeutics* 36, 143–155.

Turman, E.J., Smithson, L., Pope, L.S., Renbarger, R.E. and Stephens, D.F. (1964) Effect of feed level before and after calving on the performance of two-year-old heifers. *Agricultural Research Station Reports, Oklahoma State University Miscellaneous Publication* 74, 10–17.

Ulyatt, M.J. (1969) Progress in defining the differences in nutritive value to sheep of perennial ryegrass, short-rotation ryegrass and white clover. *Proceedings of the New Zealand Society of Animal Production* 29, 114–123.

Underwood, E.J. (1977) *Trace Elements in Human and Animal Nutrition*, 4th edn. Academic Press, New York.

Underwood, E.J. (1981) *Mineral Nutrition of Livestock*, 2nd edn. Commonwealth Agricultural Bureaux, Farnham Royal, UK.

VanDemark, N.L. and Salisbury, G.W. (1950) The relation of the post-partum breeding interval to reproductive efficiency in the dairy cow. *Journal of Animal Science* 9, 307–313.

van Rensburg, S.W.J. and de Vos, W.H. (1962) Ovulatory failure in bovines. *Onderstepoort Journal of Veterinary Research* 29, 55–79.

Vickery, P.J., Bennett, I.L. and Nicol, G.R. (1980) An improved electronic capacitance meter for estimating herbage mass. *Grass Forage Science* 35, 247–252.

Villa-Godoy, A., Hughes, T.L., Emery, R.S., Chapin, L.T. and Fogwell, R.L. (1988) Association between energy balance and luteal function in lactating dairy cows. *Journal of Dairy Science* 71, 1063–1072.

Wagner, W.C. and Hansel, W. (1969) Reproductive physiology of the post partum cow. 1. Clinical and histological findings. *Journal of Reproduction and Fertility* 18, 493–500.

Waite, R., Johnston, M.J. and Armstrong, D.G. (1964) The evaluation of artificially dried grass as a source of energy for sheep. 1. The effect of stage of maturity on the apparent digestiblity of ryegrass, cocksfoot and timothy. *Journal of Agricultural Science* 62, 391–398.

Wallace, L.R. (1955) Meeting the seasonal feed requirements of the dairy herd. In: *Proceedings of the Ruakura Farmers' Week Conference*. New Zealand Department of Agriculture, Wellington, 206–219.

Wallace, L.R. (1956) The intake and utilization of pasture by grazing dairy cattle. In: *Proceedings of the 7th International Grasslands Conference*, pp. 134–145.

Walters, D.L., Smith, M.F., Harms, P.G. and Wiltbank, J.N. (1982) Effect of steroids and/or 48 hour calf removal on serum luteinizing hormone concentrations in anestrus beef cows. *Theriogenology* 18, 349–356.

Wang, J.Y., Owen, F.G. and Larson, L.L. (1988a) Effect of beta-carotene supplementation on reproductive performance of lactating Holstein cows. *Journal of Dairy Science* 71, 181–186.

Wang, J.Y, Hafi, C.B. and Larson, L.L. (1988b) Effect of supplemental β-carotene on luteinizing hormone released in response to gonadotropin-releasing hormone challenge in ovariectomized Holstein cows. *Journal of Dairy Science* 71 498–504.

Warnick, A.C., Bedrak, E., Koger, M., Lewis, A.G. and Cunha, T.J. (1955) Reproduction performance of cattle grazing grass versus clover-grass pasture as influenced by cottonseed meal supplementation. Journal of Animal Science 14, 1259–1260.

Weigelt, B., Weigelt, R., Barth, T., Bach, S., Eulenberger, K. and Schultz, J. (1988) [Studies into embryonic mortality in a dairy herd.] *Monatshefte fur Veterinarmedizin* 43, 157–160 (in German).

Weinmann, H. (1961) Total available carbohydrates in grasses and legumes. *Herbage Abstracts* 31, 255–261.

Weston, R.H. (1982) Animal factors affecting feed intake. In: Hacker, J.B. (ed.) *Nutritional Limits to Animal Production from Pastures.* Commonwealth Agricultural Bureaux, Farnham Royal, UK, pp. 183–198.

Weston, R.H. and Hogan, J.P. (1968a) The digestion of pasture plants by sheep. I. Ruminal production of volatile fatty acids by sheep offered diets of ryegrass and forage oats. *Australian Journal of Agricultural Research* 19, 419–432.

Weston, R.H. and Hogan, J.P. (1968b) The digestion of pasture plants by sheep. II. The digestion of ryegrass at different stages of maturity. *Australian Journal of Agricultural Research* 19, 963–979.

Wetteman, R.P., Lusby, K.S. and Turman E.J. (1982) Relationship between changes in prepartum weight and condition and reproductive performance of range cows. *Agricultural Research Station Reports, Oklahoma State University Miscellaneous Publication* 112, 12–15.

Whitman, R.W. (1977) Weight change, body condition and beef-cow reproduction. *Dissertation Abstracts, International B* 37, 3289–3290.

Whittingham, D.G. and Biggers, J.D. (1967) Fallopian tube and early cleavage in the mouse. *Nature (London)* 213, 942–943.

Wilkinson, J.M. (1980) The use of animal excreta as feeds for livestock. In: *By-Products and Wastes in Animal Feeding. Occasional Publication No. 3.* British Society of Animal Production, Reading, pp. 45–60.

Willson, R.L. (1983) Free radical protection: Why vitamin E, not vitamin C, β-carotene or glutathione? In: Porter, R. and Whelan, J. (eds) *Biology of Vitamin E.* Pitman, London, pp. 19–44.

Wilson, G.F., Mackenzie, D.D.S and Holmes, C.W. (1985) Blood metabolites and infertility in dairy cows. *Proceedings of the New Zealand Society of Animal Production* 45, 17–20.

Wilson, J.G. (1952) Herd functional infertility, with reference to nutrition and mineral intake. *Veterinary Record* 64, 621–623.

Wilson, J.G. (1965) Bovine infertility – response to manganese therapy. *Veterinary Record* 77, 489–490.

Wilson, J.G. (1966) Bovine functional infertility in Devon and Cornwall: response to manganese therapy. *Veterinary Record* 79, 562–566.

Wilson, R.K. and McCarrick, R.B. (1967) A nutritional study of grass swards at progressive stages of maturity. 1. The digestibility, intake, yield and chemical composition of dried grass harvested from swards of Irish perennial ryegrass, timothy and a mixed sward at nine progressive stages of growth. *Irish Journal of Agricultural Research* 6, 267–279.

Wilson, R.K., Spillane, T.A. and Clancy, M.J. (1966) The influence of fibre content on herbage intakes by ruminants. *Irish Journal of Agricultural Research* 5, 142–143.

Wiltbank, J.N., Rowden, W.W., Ingalls, J.E., Gregory, K.E. and Koch, R.M. (1962) Effect of energy level on reproductive phenomena of mature Hereford cows. *Journal of Animal Science* 21, 219–225.

Wiltbank, J.N., Rowden, W.W., Ingalls, J.E. and Zimmerman, D.R. (1964) Influence of post-partum energy level on reproductive performance of Hereford cows restricted in energy intake prior to calving. *Journal of Animal Science* 23, 1049–1053.

Wiltbank, J.N., Bond, J., Warwick, E.J., Davis, R.E., Cook A.C., Reynolds, W.L. and Hazen, M.W. (1965) Influence of total feed and protein intake on reproductive performance in the beef female through second calving. *United States Department of Agriculture, Technical Bulletin* No. 1314.

Winks, L., Laing, A.R. and Stokoe, J. (1972) Level of urea for grazing yearling cattle during the dry season in tropical Queensland. *Proceedings of the Australian Society of Animal Production* 9, 258–261.

Witt, H.G., Warnick, A.C., Koger, M. and Cunha, T.J. (1958) The effect of level of protein intake and alfalfa meal on reproduction and gains in beef cows. *Journal of Animal Science* 17, 1211.

Wright, I.A., Rhind, S.M. and Whyte, T.K. (1992a) A note on the effects of pattern of food intake and body condition on the duration of the post-partum anoestrous period and LH profile in beef cows. *Animal Production* 54, 143–146.

Wright, I.A., Rhind, S.M., Whyte, T.K. and Smith, A.J. (1992b) Effects of body condition at calving on LH profiles and the duration of the post-partum anoestrous period in beef cows. *Animal Production* 55, 41–46.

Youdan, P.G. and King, J.O.L. (1977) The effects of body weight changes on fertility during the post-partum period in dairy cows. *British Veterinary Journal* 133, 635–641.

Young, J.S. (1965) Infertility in range cattle. *New Zealand Veterinary Journal*, 13, 1–10.

Young, J.S. (1968) Breeding patterns in commercial beef herds. 1. Herd performance in New South Wales. *Australian Veterinary Journal* 44, 350–356.

Index

abortion 49, 59
acetate
    appetite signal 21
    energy substrate 17, 31, 54
    reproduction 57
    ruminal fermentation 30
acetoacetate 31
acetyl CoA 31
adenosine diphosphate (ADP) 30, 50
adenosine triphosphate (ATP) 17, 30, 50, 55
adenosyl cobamide enzyme 48
adenyl cyclase 16
albumin 94
alfalfa 29
α-tocopherol 32, 59, 60
ambient temperature 22, 25
amino acids 17, 46, 54, 58
    glucogenic 30
    ketogenic 31
    see also isoleucine, leucine, lysine, methionine, valine
ammonia 32
anoestrus 49
anti-oxidants 39
    see also α-tocopherol, β-carotene, copper, selenium
appetite, determinants 21

barley 29
β-carotene
    anti-oxidant 59
    feed 28
    reproduction 60
β-endorphin 46
β-hydroxybutyrate 31
    energy deficiency 54
    metabolism 31
biotin 30
bodyweight see liveweight
bovine somatotrophin (BST) 46, 51
    see also growth hormone (GH)
breeding records see records, breeding
butyrate
    appetite signal 21
    energy substrate 31, 54
    ruminal fermentation 30

calcium 18, 49
calving-to-conception interval 4
calving-to-first-oestrus interval 4, 18
carbohydrate
    readily available 22, 24
    see also energy
cereals 29
ceruloplasmin 48
chou moellier 29

clover 27, 37, 88
cobalt 30, 32, 47
coenzyme B12 48
condition
    effect on reproduction 40, 44
    score 40, 45, 82
copper 32, 37, 47
    anti-oxidant 49, 60
    energy metabolism 30, 48
copper superoxide dismutase 48, 59
cottonseed meal 29
Cruciferae 29
cyanocobalamin 30, 48
cytochrome oxidase 48

deformity, congenital 49
2-deoxy-D-glucose 54, 55
digestibility 22, 32
digestible organic matter (DOM) 19, 21, 31
dopamine-β-hydroxylase 48
dry matter (DM)
    intake, cows 27
    yield, pasture 27

embryo
    death 18, 55, 59
    development 39
    metabolism 17, 31
endometrium 15
energy
    effect on reproduction 50–58
    metabolism 30–31
    see also carbohydrate, fat, infertility, protein

'fat cow syndrome' 45
fats 92
    see also energy
fatty acids 30
    acetate 32, 57
    butyrate 30
    free 17, 18, 31, 57
    polyunsaturated 59
    propionate 30, 55–60
    volatile 30, 32, 54
'fatty liver syndrome' 45

fertility, definitions 1, 6
    standards 8
    see also infertility; records; reproductive performance
fertilization 18
fetus, death 18, 47, 49
    metabolism 17, 31
fluorine 38
follicle-stimulating hormone (FSH)
    infertility 55
    physiology 15

gluconeogenesis 30, 48
glucose 47–58
    appetite signal 21
    blood concentration 22, 31, 51
    metabolism 17, 31
    requirement
      nervous function 17
      ovum, embryo, fetus 17, 46
glutathione peroxidase 59
glycerol 30
glycolysis 30
goitrogens 37, 49
gonadotrophin, pre-ovulatory surge 15
gonadotrophin-releasing hormone (GnRH)
    in energy deficiency 55, 57
    physiology 15
    treatment 95
growth hormone (GH) 22, 54
growth rate 59

hCG 95
heredity, effect on liveweight change 45, 95
hormones, plasma concentrations
    see individual entries
hypoglycaemia 48, 54–58

infertility
    definitions, 6, 35
    effects of
      age 36
      condition 40, 44–46
      energy deficiency 39

infertility *continued*
  effects of *continued*
    growth rate 36, 38, 40
    hypoglycaemia 54, 57
    liveweight 36
    milk yield 46
    mineral deficiency 37, 39, 47
    non-nutritional disease 39
    pasture/crop maturity 36
    plane of nutrition 35, 38, 41–43
    poisons 37, 38
    protein deficiency 38, 57
    vitamin deficiency 39, 60
inhibin 17
insulin 17, 22, 54
intake 21
  measurement of 86, 93
intervals
  calving-to-conception 4
  calving-to-first oestrus 4, 18, 51
  calving-to-mating 4, 5
  inter-calving 4, 5, 46
  inter-oestrus 18, 59
iodine 32, 47, 49, 55
iron 47
isoleucine 58

kale 29, 37, 50
ketosis 40, 54

lactate 17, 46
lactation 18
  effect on reproduction 46, 50
leucine 58
lignin 22
linseed meal 29
liveweight, effect on fertility 40, 44, 61
  targets 93
  see also condition
lucerne 29, 37
luteinizing hormone (LH)
  infertility 51, 54, 57
  physiology 15–18
lysine 58
lysyloxidase 48

magnesium 18, 30, 50

maize silage 37, 60
manganese 30, 32, 37, 47, 49
mangolds 29
metabolism, reproductive system 17
metabolites, normal blood concentrations 95
methionine 37, 58
minerals *see* cobalt, copper, iodine, iron, magnesium, manganese, molybdenum, phosphorus, sulfur, selenium
molybdenum 47, 60
monensin 55
monoamine oxidase 48

nicotinamide 31
nitrate 25, 32, 58
nitrogen
  N:ME ratio 32
  see also amino acids, non-protein nitrogen, protein
nitrogenous fertilizer 24
non-esterified fatty acids 54
non-protein nitrogen 22, 31, 57

oats, forage 23, 29
oestradiol-17β 16, 55
oestrogen 15, 54
oestrous cycle 15, 46, 52
oestrus 15, 18, 49, 55
oocyte 16
opioid peptides 17, 46, 57
ovulation 15, 49
ovum development 15, 18
  death 60
oxaloacetate 17, 31
oxytocin 16

pantothenic acid 30
parturition syndrome 40
Paspalum dilatatum 37
pasture 20, 27
  composition 22–28
  fertilizer treatment 24, 31
  intake 27
  maturity 23, 27
  species 24, 27
  yield 27, 86

peanut meal 29
pH, soil 49, 90
phlorizin 54
phosphoenolpyruvate carboxyl kinase 49
phosphoglucomutase 49
phosphorus
    effect on reproduction 47, 49, 50, 54
    metabolism 30–32
phyto-oestrogens 37
plane of nutrition, effect on reproduction 40–43
poisons 76
pollution 76
potassium 18, 30, 50
pregnancy 16, 46
    rates 7, 8, 46, 49, 51, 59
pregnant mare's serum gonadotrophin (PMSG) 95
pregnenolone 16
progesterone
    effect of energy deficiency 54
    physiology 15–17
    therapy 95
progesterone-impregnated intra-vaginal device 95
propionate 30, 50–55
    appetite signal 21
prostaglandin $PGF_{2\alpha}$ 16
protein 31, 57
    see also energy, non-protein nitrogen
puberty 40, 52, 61
pyridoxine 30
pyruvate carboxylase 49

rain 25
rape 29
records, breeding 8
    analysis 9, 64–71
    differential diagnosis 73
    examples 10–13, 73, 77–81
    interpretation 71
    monitoring 8
      extensively managed herds 15, 95
      intensively managed herds 9–14
reproductive performance
    standards 7
    see also fertility, infertility
retained fetal membranes 49, 59
riboflavin 30
rumen, fermentation 30
ryegrass 23, 27

safflower meal 29
selenium 32, 37, 47, 59
sodium 18, 50
somatotrophin 46, 51
sorghum 29
soyabean meal 30
starch 30
straw 29
stocking rate 27, 35
suckling 18, 46
sugarbeet 29, 50
sulfur 47, 54
sunlight 25

terminal oxidative pathway 17, 30
thiamine 30
thiocyanates 49
thyroxine 49
tricarboxylic acid cycle 17, 30
trophoblast protein-1 (TP-1) 16
turnips 29

urea 58

valine 58
vitamins 39, 59

water
    drinking 22
    plant intracellular 22
wheat 29

yield
    milk 27, 28
    pasture 27